21 Culinary Herbs

Roby Jose Ciju

DEDICATION

This book is dedicated to all home gardeners, farmers, commercial growers and all plant-loving souls who have a genuine interest in growing plants...

Table of Contents

Spinach

Spinach is a very popular leafy vegetable in many parts of the world. Scientific name of spinach is *Spinacia oleracea* and it belongs to the family Chenopodiaceae. However some people place this plant in the family Amaranthacea. Spinach is also known by the names 'true spinach' and 'English spinach'.

Spinach is mainly grown for its edible leaves which are considered as a highly nutritious vegetable. The edible part consists of a compact rosette of bright green leaves.

Spinacia oleracea is a multistem herbaceous vegetable grown as an annual crop. Optimum temperature requirement for its growth is 13-20 degree Celsius. Optimum annual rainfall requirement is 800-1200 mm. Spinach can be grown at an altitude of 3600 meters from MSL (mean sea level).

Origin and Distribution: Spinach is believed to be originated in South-West Asia.

Taxonomy: A detailed taxonomic classification of spinach is as given below:

Kingdom	Plantae – Plants
Subkingdom	Tracheobionta – Vascular plants
Superdivision	Spermatophyta – Seed plants
Division	Magnoliophyta – Flowering plants
Class	Magnoliopsida – Dicotyledons
Subclass	Caryophyllidae
Order	Caryophyllales
Family	Chenopodiaceae – Goosefoot family
Genus	Spinacia L. – spinach
Species	Spinacia oleracea L. – spinach

Botanical Description: A detailed botanical description of spinach is as given below:

Stems	Spinach has a leafy stem.
Leaves	Leaves are bright green and oblong shaped; usually wider than long
Flowering	Flowering time is spring-fall
Seeds	Spinach has tiny, spiny or prickly seeds

Food Uses: Spinach is a highly nutritious vegetable rich in vitamins, calcium, iron, and antioxidant carotenoids. Tender and succulent leaves of spinach are used in salad preparations, soups and omelettes. Spinach may be cooked as a main vegetable dish or may be used as a side ingredient along with other vegetables. Spinach is considered as one of the highly nutritious leafy vegetables that are available today for human consumption. However over-consumption of spinach must be avoided due to the presence of oxalates in the leaves. Nutritive value of 100 grams of edible portion of raw spinach is given below:

Nutrient	Unit	Value per100g
Water	g	91.4
Energy	kcal	23
Protein	g	2.86
Total lipid (fat)	g	0.39
Carbohydrate, by difference	g	3.63
Fiber, total dietary	g	2.2
Sugars, total	g	0.42
Calcium, Ca	mg	99
Iron, Fe	mg	2.71
Magnesium, Mg	mg	79
Phosphorus, P	mg	49
Potassium, K	mg	558
Sodium, Na	mg	79
Zinc, Zn	mg	0.53
Vitamin C, total ascorbic acid	mg	28.1
Thiamin	mg	0.078
Riboflavin	mg	0.189
Niacin	mg	0.724
Vitamin B-6	mg	0.195
Folate, DFE	Âµg	194
Vitamin B-12	Âµg	0
Vitamin A, RAE	Âµg	469
Vitamin A, IU	IU	9377
Vitamin E (alpha-tocopherol)	mg	2.03
Vitamin K (phylloquinone)	Âµg	482.9
Lipids		
Fatty acids, total saturated	g	0.063
Fatty acids, total monounsaturated	g	0.01
Fatty acids, total polyunsaturated	g	0.165

Nutrition in Cooked Spinach: Spinach is cooked as a leafy vegetable, boiled as soups and may be blanched to use as a salad green. Nutrition in 100 grams of edible portions of cooked spinach is given below:

Nutrient	Unit	Value per100g
Water	g	91.21
Energy	kcal	23
Protein	g	2.97
Total lipid (fat)	g	0.26
Carbohydrate, by difference	g	3.75
Fiber, total dietary	g	2.4
Sugars, total	g	0.43
Calcium, Ca	mg	136
Iron, Fe	mg	3.57
Magnesium, Mg	mg	87
Phosphorus, P	mg	56
Potassium, K	mg	466
Sodium, Na	mg	70
Zinc, Zn	mg	0.76
Vitamin C, total ascorbic acid	mg	9.8
Thiamin	mg	0.095
Riboflavin	mg	0.236
Niacin	mg	0.49
Vitamin B-6	mg	0.242
Folate, DFE	µg	146
Vitamin B-12	µg	0
Vitamin A, IU	IU	10481
Vitamin E (alpha-tocopherol)	mg	2.08
Vitamin K (phylloquinone)	µg	493.6
Fatty acids, total saturated	g	0.043
Fatty acids, total monounsaturated	g	0.006
Fatty acids, total polyunsaturated	g	0.109

Nutrition in Frozen Spinach: Nutrition in 100 grams of Frozen Spinach (chopped or leaf, unprepared) is given below:

Nutrient	Unit	Value per100g
Water	g	90.17
Energy	kcal	29
Protein	g	3.63
Total lipid (fat)	g	0.57
Carbohydrate, by difference	g	4.21
Fiber, total dietary	g	2.9
Sugars, total	g	0.65
Calcium, Ca	mg	129
Iron, Fe	mg	1.89
Magnesium, Mg	mg	75
Phosphorus, P	mg	49
Potassium, K	mg	346
Sodium, Na	mg	74
Zinc, Zn	mg	0.56
Vitamin C, total ascorbic acid	mg	5.5
Thiamin	mg	0.094
Riboflavin	mg	0.224
Niacin	mg	0.507
Vitamin B-6	mg	0.172
Folate, DFE	µg	145
Vitamin B-12	µg	0
Vitamin A, RAE	µg	586
Vitamin A, IU	IU	11726
Vitamin E (alpha-tocopherol)	mg	2.9
Vitamin D	IU	0
Vitamin K (phylloquinone)	µg	372
Fatty acids, total saturated	g	0.041
Fatty acids, total monounsaturated	g	0
Fatty acids, total polyunsaturated	g	0.082

Nutrition in 100 grams of *cooked* Frozen Spinach (chopped or leaf, cooked, boiled, drained, without salt) is given below:

Nutrient	Unit	Value per100g
Water	g	88.94
Energy	kcal	34
Protein	g	4.01
Total lipid (fat)	g	0.87
Carbohydrate, by difference	g	4.8
Fiber, total dietary	g	3.7
Sugars, total	g	0.51
Calcium, Ca	mg	153
Iron, Fe	mg	1.96
Magnesium, Mg	mg	82
Phosphorus, P	mg	50
Potassium, K	mg	302
Sodium, Na	mg	97
Zinc, Zn	mg	0.49
Vitamin C, total ascorbic acid	mg	2.2
Thiamin	mg	0.078
Riboflavin	mg	0.176
Niacin	mg	0.439
Vitamin B-6	mg	0.136
Folate, DFE	Âµg	121
Vitamin B-12	Âµg	0
Vitamin A, IU	IU	12061
Vitamin E (alpha-tocopherol)	mg	3.54
Vitamin K (phylloquinone)	Âµg	540.7
Fatty acids, total saturated	g	0.157
Fatty acids, total monounsaturated	g	0
Fatty acids, total polyunsaturated	g	0.371

Growing Practices for Spinach: Spinach varieties suitable for cultivation should be selected depending upon growing purposes. There are smooth-leaved varieties and savoy-leaved varieties. '*Virginia Savoy*' is a prominent savoy-leaved variety. '*Early Smooth Leaf*' is a smooth-leaved variety. A comparative study between savoy-leaved and smooth-leaved spinach varieties is given below:

Savoy-leaved Spinach	Smooth-leaved Spinach
Leaves are blistered, large, and dark-green in colour	Leaves are smooth, smaller, and light green in colour and with a pointed apex
Seeds are prickly or spiny	Seeds are smooth and round

In USA, major commercial varieties are Imperial Spring, Shasta, Polka, Spectrum, Sporter, Bossanova, Spark and Space. According to a study conducted at UC Davis, varieties Imperial Spring, Shasta, Polka, Spectrum and Sporter have notably longer shelf- life in storage than do varieties Bossanova, Spark and Space.

Climatic Requirements: Spinach is suitable for growing in semiarid, subtropical humid, subtropical dry summer, subtropical dry winter, temperate oceanic, temperate continental, temperate with humid winters, and temperate with dry winters climate zones.

Sunlight: Light intensity should be very bright. Clear skies are preferred.

Soil Requirements: Optimum soil depth preferred is 20-50 cm. Spinach may be grown in all types of soils but well-drained loamy soils are the most ideal. In other words, soil texture preferred for its growth is medium or light soils. Soil fertility should be high. Soil should be well-drained. Soil salinity should be low. Optimum soil pH is 6-7.5.

Propagation: Propagation is via seeds. Since spinach is a cross-pollinated crop, it is difficult to produce true-to-type seeds for small-scale cultivations. Good quality hybrid seeds that are available from certified nurseries may be used for propagating spinach plants.

Sowing Seeds: Seeds are directly sown in the well-prepared main field by broadcasting. Approximately 30 kilograms of seeds are required to sow one hectare of land. For smaller areas, raised beds or ridges may be prepared in the land and seeds are sown in rows at a distance of 15-20 cm apart. For container growing, seedlings may be raised indoors and later transplanted in containers when seedlings are about 15-20 cm tall.

Sowing Time: Spinach may be grown two or three times in a year. Seeds may be sown in early spring (March to May), at the onset of monsoons and at the onset of winter in tropics (i.e. September to November). For container growing, seedlings may be raised at any time of the year according to the convenience of the grower.

Watering: Moist soil is preferred for spinach growth. On average spinach plants need to be watered at weekly intervals. While watering the plants, make sure that top 15-20 cm of soil is watered properly. After every harvest, watering is recommended to boost leaf production.

Fertilizer and Manure Application: Since spinach is a leafy vegetable, nitrogen fertilizers must be added in plenty to boost leaf growth. Approximately 40 tons of farm yard manure or compost per hectare of land may be incorporated into the top soil at the time of field preparation. After that a top dressing of 20 kilograms of nitrogen per hectare may be given after every harvest. For container growing, foliar application of a liquid nitrogen fertilizer is a good idea after every harvesting of leaves.

Harvesting and Yield: Crop is ready for harvest within two to three months after sowing. Tender, succulent leaves are cut from the base by using a sharp blade or knife when they are about 25-30 cm long. Several harvestings are possible depending upon the growth of the plant. Approximately 8-10 tons per hectare yield is obtained under best cultural practices. In other words, approximately 4000 to 5000 Kg spinach leaves are obtained from one acre of land.

Disease Management: Major fungal infections of spinach are damping-off, Cercospora leaf spot, mildew and rust. Damping-off is a fungal infection of young seedlings and can be controlled by treating the seeds with a recommended fungicide before sowing. Cercospora leaf spot produces black, dead spots on leaves and this fungal infection can be controlled by restricted use of fungicide Bordeaux mixture. Mildew and rust can be controlled by sulphur dusting.

Insect-Pest Management: There are no major insect-pest infestations found in spinach crop. Sometimes, some aphids, caterpillars or beetles may attack the crop but this can effectively be controlled by organic pesticides such as dusting with 0.2-0.3% pyrethrum or by spraying tobacco emulsion. Since it is a leafy vegetable, application of chemical insecticides is not recommended for insect management.

Seed Production in Spinach: Spinach is a cross-pollinated crop and it is pollinated by wind. There are four different types of plant in spinach. These are extreme males, vegetative males, females, and hermaphrodite. Seed stalks first emerge in extreme males followed by vegetative males. These seed stalks bear male flowers but do not produce seed. Seed stalks of hermaphrodite and female plants appear later. Separate male and female plants are used for commercial production of hybrid seeds. Hybrid seeds give more yield and better quality greens.

Post Harvest Management of Spinach: A detailed description of post harvest management practices for spinach crop is given below:

Maturity Indices	Young, tender, clean and succulent leaves that are midway to maturity are harvested for fresh markets. Older and yellowing leaves are to be avoided while harvesting. Generally 20-30 days of re-growth is required before second harvest.
Quality Indices	Leaves should be uniformly green without any yellowing. Leaves should be fully succulent, clean, and free from bruises and injuries. For bunched spinach, roots should be trimmed from the base.
U.S. Grades	Bunched — U.S. No. 1, No. 2 Leaves — U.S. Extra No. 1, No. 1, Commercial
Optimum Storage Temperature	All leafy vegetables are highly perishable. At 0°C (32°F), Spinach may be stored up to 2 weeks without losing its freshness.
Optimum Relative Humidity	95-98%
Responses to Ethylene	Spinach is highly sensitive to ethylene and hence yellowing of leaves is accelerated even at low levels of ethylene. During distribution and short-term storage, probable ethylene exposure should be avoided by not mixing spinach loads with ethylene producing items such as apples, melons and tomatoes.
Physiological and Physical Disorders	*Freezing Injury*: Freezing injury will be initiated at -0.3°C (31.5°F) which will be resulted in soft rot. Symptoms are watersoaking followed by rapid decay of leaves. *Yellowing*: Yellowing of leaves due to ethylene presence in storage or by the presence of multiple injuries and bruises to leaves and petioles.
Pathological Disorders	Bacterial Soft-Rot caused by *Erwinia* and *Pseudomonas*.
Other Considerations	Package-icing and frequent light misting may be done to delay wilting and enhance shelf life.

Malabar Spinach or Ceylon Spinach

Malabar spinach is not a true spinach. Its leaves resemble spinach leaves and edible properties are almost similar to true spinach and hence the name 'Malabar spinach'. There are green-stemmed (Green Malabar) and red-stemmed (Red Malabar) varieties of Malabar spinach. Botanical name of green Malabar spinach is *Basella alba* and Synonym is *Basella cordifolia*. Botanical name of red Malabar spinach is *Basella rubra*.

Malabar spinach belongs to family Basellaceae. It is a fast-growing herbaceous perennial plant and is edible. The leaves and tender shoots of the plant are used as a leafy vegetable just like spinach. Even though it is a perennial plant it is grown as an annual for food purposes.

Common Names : Other common names of Malabar spinach are Basella, Cambian spinach, Indian spinach, vine spinach, Malabar nightshade, country spinach, bertalha vermelha, Malabarspinat, basela branca, bertalha branca, basela vermelha, melao de soldado, sabao de soldado, Baselle, Brede de malebar, Espinaca, Espinaca de celilan, Poi, Pui, Alugbati, Grana, Remayong, Gendola, Bretana, Libato, Acelca trepadora, Hung tang, Tsoi, Shaan tsoi, Genjerot, Jingga, Tsuru murasaki, Remayong, Niviti, Pasali, Pak prang, Climbing spinach, Creeping spinach, buffalo spinach, Surinam spinach, Chinese spinach, Vietnamese spinach and broad bologi.

Red-stemmed variety of Malabar spinach is also known as red stem Malabar spinach, and red vine spinach. Similarly, green-stemmed variety is known as white stem Malabar spinach, and white vine spinach.

Taxonomy

Kingdom	Plantae – Plants
Subkingdom	Tracheobionta – Vascular plants
Superdivision	Spermatophyta – Seed plants
Division	Magnoliophyta – Flowering plants
Class	Magnoliopsida – Dicotyledons
Subclass	Caryophyllidae
Order	Caryophyllales
Family	Basellaceae – Basella family
Genus	Basella L. – Basella
Species	Basella alba L. – Ceylon spinach

Origin: Malabar spinach is found growing abundantly in tropical continents such as Asia and Africa. It is believed to be originated in the tropical regions of Malabar Coast of India and Sri Lanka.

Botanical Description

Plant	The plant has vining habit, loves warm-season, fast-growing, and soft-stemmed. Vines reach up to 10 metres in length upon full growth.
Stems	Stems are trailing vines with short-petioled leaves and small, black fruits. Stems are green in alba variety and red stems in rubra variety
Leaves	Leaves are glossy, dark green, thick, succulent, savoyed and heart-shaped with a flavour
Fruits	Fruit is a black-coloured berry which may be ground and used as a food colouring agent
Flowering	Malabar spinach does not flower in day lengths longer than 13 hours. In other words, flowering is induced in short-day months (Nov-Feb) in tropics

Growth Habit: Malabar spinach may be trained on trellises as a spreading vine crop or trained into a bush with proper care and maintenance.

Food Uses: Malabar spinach is used like spinach in salad greens, and cooking as a leafy vegetable. Leaves are dark green and thick with a mucilaginous texture which is rich in dietary fiber. Hence leaves may be boiled to thicken the consistency of soups. Malabar spinach may be cooked as stir-fries or as a main vegetable dish or as an additional ingredient in other vegetable preparations or as an ingredient in meat and fish preparations. Raw Malabar spinach is rich source of calcium, potassium, vitamin A and vitamin C. Nutrition in 100 grams of edible portions of raw Malabar spinach (vine spinach) is given below:

Nutrition in Raw Vine Spinach (Basella): Nutrition in 100 grams of edible portion of RAW Malabar spinach is given below:

Nutrient	Unit	Value per100g
Water	g	93.1
Energy	kcal	19
Protein	g	1.8
Total lipid (fat)	g	0.3
Carbohydrate, by difference	g	3.4
Calcium, Ca	mg	109
Iron, Fe	mg	1.2
Magnesium, Mg	mg	65
Phosphorus, P	mg	52
Potassium, K	mg	510
Sodium, Na	mg	24
Zinc, Zn	mg	0.43
Vitamin C, total ascorbic acid	mg	102
Thiamin	mg	0.05
Riboflavin	mg	0.155
Niacin	mg	0.5
Vitamin B-6	mg	0.24
Folate, DFE	µg	140
Vitamin B-12	µg	0
Vitamin A, IU	IU	8000

Nutrition in Cooked Malabar Spinach: Nutrition in 100 grams of edible portion of cooked Malabar spinach is given below:

Nutrient	Unit	Value per100g
Water	g	92.5
Energy	kcal	23
Protein	g	2.98
Total lipid (fat)	g	0.78
Carbohydrate, by difference	g	2.71
Fiber, total dietary	g	2.1
Calcium, Ca	mg	124
Iron, Fe	mg	1.48
Magnesium, Mg	mg	48
Phosphorus, P	mg	36
Potassium, K	mg	256
Sodium, Na	mg	55
Zinc, Zn	mg	0.3
Vitamin C, total ascorbic acid	mg	5.9
Thiamin	mg	0.106
Riboflavin	mg	0.129
Niacin	mg	0.787
Vitamin B-6	mg	0.086
Folate, DFE	Âµg	114
Vitamin B-12	Âµg	0
Vitamin A, IU	IU	1158

Growing Practices for Malabar Spinach: Malabar spinach is a warm season crop and grows well in hot, humid climates. Growth of this plant is generally limited at altitudes greater than 500 metres (1,600 ft) above sea level. In other words, the plant grows well at altitudes lower than 500 metres from sea level.

Sunlight: Full sunlight and hot weather are favoured by Malabar spinach plants.

Soil Requirements: Even though Malabar spinach grows well in a wide range of soils, the most ideal soil is moist, fertile sandy loam soils which are rich in organic matter (humus) with pH ranging from 6.5 to 6.8. Malabar spinach is highly sensitive to frost but tolerates damp wet soils.

Propagation: Malabar spinach is propagated by seeds and stem cuttings.

Site Preparation/Field Preparation: Site of planting/seeding is prepared as in case of spinach cultivation.

Sowing Method: Direct seeding or broadcasting is done as in case of spinach if seeds are used for propagation. If stem cuttings are used, they are directly planted in the pits prepared at the site.

Sowing Time/Planting Time: In warmer tropics, sowing or planting is done 2 to 3 weeks after last frost. In colder areas, seedlings are raised indoors and later transplanted in the site 2 to 3 weeks after last frost.

Seed Treatment: It may take 2-3 weeks for the seeds to emerge. Seed scarification is recommended in order to hasten germination process.

Seed Scarification: Seed coats are carefully cut through by using a sharp knife or sandpaper to speed germination.

Spacing: Transplants are planted about 30 cm apart within a row

Training: Malabar spinach is a trailing vine plant and its long trailing vines need to be trained on trellises or other support structures to facilitate harvesting and other cultural practices.

Watering: Moist soil is preferred for Malabar spinach growth. On average Malabar spinach plants need to be watered at weekly intervals. While watering the plants, make sure that top 15-20 cm of soil is watered properly. After every harvest, watering is recommended to boost leaf production.

Fertilizer and Manure Application: Since leaf production has to be promoted, application of liberal doses of nitrogen fertilizers are recommended. Farm yard manure or compost @40-50tons/ha may be incorporated into the top soil at the time of field preparation. After that a top dressing of 20 kilograms of nitrogen per hectare may be given after every harvest. For container growing, foliar application of a liquid nitrogen fertilizer is a good idea after every harvesting of leaves.

Harvesting: It may take 2-3 months for a seedling to reach harvest maturity. In case of stem cuttings, harvest maturity is reached within one to two months.

Yield: Under good cultural management a Malabar spinach crop may yield about 50 tons/ha of land.

Mustard Spinach Tendergreen or Japanese Spinach

Mustard spinach is a green leafy vegetable just like spinach and mustard leaves but it is neither mustard nor spinach. However it belongs to mustard family, i.e. Brassicaceae. In other words, it is a leafy brassica vegetable grown for its highly nutritious edible leaves. Scientific name is *Brassica perviridis* or *Brassica rapa perviridis*.

Taxonomy

Kingdom	Plantae – Plants
Subkingdom	Tracheobionta – Vascular plants
Superdivision	Spermatophyta – Seed plants
Division	Magnoliophyta – Flowering plants
Class	Magnoliopsida – Dicotyledons
Subclass	Dilleniidae
Order	Capparales
Family	Brassicaceae / Cruciferae – Mustard family
Genus	Brassica L. – mustard
Species	Brassica perviridis

Common Names: Komatsuna, Japanese mustard spinach, tender green mustard, spinach mustard

Food Uses: Mustard spinach is cooked just like spinach. It is used as salad greens and cooked as a delicious leafy vegetable. It may be boiled to soups or cooked as stir-fried. Mustard spinach is rich in vitamins and minerals. Nutrition in 100 grams of edible portion of raw mustard spinach is given below:

Nutrient	Unit	Value per100g
Water	g	92.2
Energy	kcal	22
Protein	g	2.2
Total lipid (fat)	g	0.3
Carbohydrate, by difference	g	3.9
Fiber, total dietary	g	2.8
Calcium, Ca	mg	210
Iron, Fe	mg	1.5
Magnesium, Mg	mg	11
Phosphorus, P	mg	28
Potassium, K	mg	449
Sodium, Na	mg	21
Zinc, Zn	mg	0.17
Vitamin C, total ascorbic acid	mg	130
Thiamin	mg	0.068
Riboflavin	mg	0.093
Niacin	mg	0.678
Vitamin B-6	mg	0.153
Folate, DFE	Âµg	159
Vitamin B-12	Âµg	0
Vitamin A, IU	IU	9900
Fatty acids, total saturated	g	0.015
Fatty acids, total monounsaturated	g	0.138
Fatty acids, total polyunsaturated	g	0.057

Nutrition in Raw Mustard Spinach (tendergreen); cooked, boiled, drained, without salt

Nutrient	Unit	Value per100g
Water	g	94.5
Energy	kcal	16
Protein	g	1.7
Total lipid (fat)	g	0.2
Carbohydrate, by difference	g	2.8
Fiber, total dietary	g	2
Calcium, Ca	mg	158
Iron, Fe	mg	0.8
Magnesium, Mg	mg	7
Phosphorus, P	mg	18
Potassium, K	mg	285
Sodium, Na	mg	14
Zinc, Zn	mg	0.11
Vitamin C, total ascorbic acid	mg	65
Thiamin	mg	0.041
Riboflavin	mg	0.062
Niacin	mg	0.43
Vitamin B-6	mg	0.097
Folate, DFE	µg	73
Vitamin B-12	µg	0
Vitamin A, IU	IU	8200

Growing Practices for Mustard Spinach: Mustard spinach thrives well in cool climate even though it can be grown in warmer climates also. Propagation is via seeds Seeds may be sown almost a month before last frost. It may take 5-10 days for seedlings to emerge. Direct seeding in groups of 3 seeds is practiced if it is grown in open fields. For container growing of mustard spinach, seedlings are raised indoors. Seeds are sown at a depth of 0.5 to 1 cm

Site/Filed Preparation: Site is prepared by ploughing and levelling the land. Raised beds or ridges are prepared and later pits are made on these ridges to sow the seeds.

Seed Spacing: Seeds are sown in pits which are spaced at 15-20 cm apart within a row

Row Spacing: Row to row distance within a ridge is about 2.5 -5 cm

Thinning: When plants are 10-15 cm tall, thinning is done keeping one plant for every 15-20 cm distance within a row.

Watering: Moist soil is preferred for mustard spinach growth. Soon after sowing seeds and thinning a light irrigation is given. On average mustard spinach plants need to be watered at weekly intervals. While watering the plants, make sure that top 15-20 cm of soil is watered properly. After every harvest, watering is recommended to boost leaf production.

Fertilizer and Manure Application: Since leaf production has to be promoted, application of liberal doses of nitrogen fertilizers are recommended. Farm yard manure or compost @40-50tons/ha may be incorporated into the top soil at the time of field preparation. After that a top dressing of 20 kilograms of nitrogen per hectare may be given after every harvest. For container growing, foliar application of a liquid nitrogen fertilizer is a good idea after every harvesting of leaves.

Harvesting: Growth duration of mustard spinach is approximately one or two months. Harvesting is done when plants are about 20-30 cm tall. Harvesting is done with a sharp knife or blade from the outside of the plants while interior leaves are left undisturbed to continue growth. Harvesting may be done at weekly or biweekly intervals as and when new growth appears. Continuous harvesting is possible only when harvesting is done from outside towards interior of the plant while leaving new growth undisturbed. For salad uses, tender plants may carefully be pulled from the base and roots are trimmed.

New Zealand Spinach

Scientific name of New Zealand spinach is *Tetragonia tetragonioides* and it belongs to the family Aizoaceae. Scientific synonyms are *T. expansa* and *Demidovia tetragonioides*. *Tetragonia tetragonioides* or New Zealand spinach is a dicot plant grown for its edible leaves. The plant has a dense, spreading growth habit with its trailing stems have a tendency to form ground cover if its growth is unchecked. However, for vegetable purposes, it is grown and trained as an herbaceous annual bushy plant. It is perennial in its natural growth habit.

Botanical Description

Plant	It is a low growing, vigorous, spreading and much-branched herbaceous plant; reaches up to 30-40 cm in height upon maturity with a spread of about one meter
Stems	Trailing stems which are mat-forming
Leaves	Leaves are 3–15 cm long, thick, succulent, and triangular-shaped; it is bright green in colour; leaves have a lettuce-like flavour.
Inflorescences and flowers	Yellowish flowers are borne on 2 mm long peduncles
Fruits	Fruit is a small, hard pod of 8-12 mm length which are covered with 4-6 small horns
Seeds	Smooth irregularly-shaped seeds
Flowering	Flowering is during spring-fall

Common Names: Common names of New Zealand spinach are Warrigal cabbage, Tetragone, Tetragone cornue, Epinard d'ete, Ispinaka, Extranjera de Nueva Zelandia, Tetragono, Neuseelander Spinat, Nieuwzeelandse spinazie, Espinaca de New Zealand, Kabak, tetragonia, ice plant, everbearing spinach, everlasting spinach, perpetual spinach, Della Nuova Zelanda, Warrigal greens, sea spinach, Botany Bay spinach, tetragon and Cook's cabbage.

Origin and Distribution: New Zealand spinach is believed to be a native of coastal areas of Australia, Tasmania, New Zealand, China, Taiwan, southern South America, and some Pacific Islands, including Japan.

Taxonomy

Kingdom	Plantae – Plants
Subkingdom	Tracheobionta – Vascular plants
Superdivision	Spermatophyta – Seed plants
Division	Magnoliophyta – Flowering plants
Class	Magnoliopsida – Dicotyledons
Subclass	Caryophyllidae
Order	Caryophyllales
Family	Aizoaceae – Fig-marigold family
Genus	Tetragonia L. – tetragonia
Species	Tetragonia tetragonioides (Pall.) Kuntze – New Zealand spinach

Food Uses: New Zealand spinach is used and cooked like other spinach herbs. It is used in salad greens, soups and vegetable dishes. New Zealand can be preserved by freezing just like spinach. Raw leaves are not suitable for consumption because of the presence of oxalates in them which may be allergic to some people. Blanching of leaves is done to remove oxalates just before cooking the leaves. Blanching is a process where freshly harvested leaves are immersed in hot water for a minute and then these leaves are rinsed with cold water. New Zealand spinach is cooked as a major vegetable dish or may be pickled. Blanched leaves may be used as an ingredient in salads. It may also be used in soups and stews. It may be added as an ingredient in pastas and omelettes. It is low in calories, proteins, and carbohydrates and fat, but high in nutrients.

It is an excellent source of vitamin A, with 4,400 international units (IU) per 100 gram. Other notable vitamins include 30 mg of vitamin C per 100 gram of edible portion. It is rich in minerals such as 130 milligrams each of sodium and potassium in 100 grams of edible portion and 58 milligrams of calcium, 39 milligrams of magnesium and 28 milligrams of phosphorous in 100 grams of edible portion.

Nutrition in Raw New Zealand Spinach: Nutrition in 100 grams of edible portion of RAW New Zealand spinach is given below:

Nutrient	Unit	Value per100g
Water	g	94
Energy	kcal	14
Protein	g	1.5
Total lipid (fat)	g	0.2
Carbohydrate, by difference	g	2.5
Fiber, total dietary	g	1.5
Sugars, total	g	0.29
Calcium, Ca	mg	58
Iron, Fe	mg	0.8
Magnesium, Mg	mg	39
Phosphorus, P	mg	28
Potassium, K	mg	130
Sodium, Na	mg	130
Zinc, Zn	mg	0.38
Vitamin C, total ascorbic acid	mg	30
Thiamin	mg	0.04
Riboflavin	mg	0.13
Niacin	mg	0.5
Vitamin B-6	mg	0.304
Folate, DFE	µg	15
Vitamin B-12	µg	0
Vitamin A, IU	IU	4400
Vitamin E (alpha-tocopherol)	mg	1.42
Vitamin K (phylloquinone)	µg	337
Fatty acids, total saturated	g	0.032
Fatty acids, total monounsaturated	g	0.005
Fatty acids, total polyunsaturated	g	0.084

Nutrition in Cooked New Zealand Spinach (boiled, drained, without salt): Nutrition in 100 grams of edible portion of COOKED New Zealand spinach is given below:

Nutrient	Unit	Value per100g
Water	g	94.8
Energy	kcal	12
Protein	g	1.3
Total lipid (fat)	g	0.17
Carbohydrate, by difference	g	2.13
Fiber, total dietary	g	1.4
Sugars, total	g	0.25
Calcium, Ca	mg	48
Iron, Fe	mg	0.66
Magnesium, Mg	mg	32
Phosphorus, P	mg	22
Potassium, K	mg	102
Sodium, Na	mg	107
Zinc, Zn	mg	0.31
Vitamin C, total ascorbic acid	mg	16
Thiamin	mg	0.03
Riboflavin	mg	0.107
Niacin	mg	0.39
Vitamin B-6	mg	0.237
Folate, DFE	µg	8
Vitamin B-12	µg	0
Vitamin A, IU	IU	3622
Vitamin E (alpha-tocopherol)	mg	1.23
Vitamin K (phylloquinone)	µg	292
Fatty acids, total saturated	g	0.027
Fatty acids, total monounsaturated	g	0.005
Fatty acids, total polyunsaturated	g	0.071

Growing Practices for New Zealand Spinach: The crop is best grown at an altitude of 1000-1700 meters from mean sea level. It is drought tolerant but does not like scorching heat. The plant prefers a moist environment for its growth. Flavour of leaves is well-developed when the plant is grown with consistent moisture.

Sunlight: New Zealand spinach prefers full sunlight for its healthy growth.

Soil Requirements: New Zealand spinach may be grown in a wide variety of soils. However, the most ideal soil for its healthy growth is well-drained, fertile, sandy soils, which are rich in organic matter (humus). Ideal soil pH is 6.8 to 7.0. The plant grows well in saline conditions.

Propagation: Propagation is via seeds. In warmer tropics, sowing is done 2 to 3 weeks after last frost. In colder areas, seedlings are raised indoors 2 to 3 weeks before last frost and later transplanted in the site 2 to 3 weeks after last frost. Before planting, the seeds should be soaked for 12-24 hours in cold water, or 3-5 hours in warm water. This will facilitate easy germination. It may take 10-20 days for seedlings to emerge

Sowing Method: Direct seeding in groups of 3 seeds is practiced if it is grown in open fields. For container growing, seedlings are raised indoors.

Site/Filed Preparation: Site is prepared by ploughing and levelling the land. Raised beds or ridges are prepared and later pits are made on these ridges to sow the seeds.

Seed Depth: Seeds are sown at a depth of 0.5 to 1 cm

Spacing: Seeds are sown in pits which are spaced at 15-30 cm apart within a row. Row to row distance within a ridge is about 2.5 -5 cm

Thinning: When plants are 10-15 cm tall, thinning is done keeping one plant for every 15-30 cm distance within a row. Plant to plant distance is kept more because of the spreading growing habit of the plant.

Watering: Moist soil is preferred for New Zealand spinach growth. Soon after sowing seeds and thinning a light irrigation is given. On average the plants need to be watered at weekly intervals. While watering the plants, make sure that top 15-20 cm of soil is watered properly. After every harvest, watering is recommended to boost leaf production.

Fertilizer and Manure Application: Since leaf production has to be promoted, application of liberal doses of nitrogen fertilizers are recommended. Farm yard manure or compost @40-50tons/ha may be incorporated into the top soil at the time of field preparation. After that a top dressing of 20 kilograms of nitrogen per hectare may be given after every harvest. For container growing, foliar application of a liquid nitrogen fertilizer is a good idea after every harvesting of leaves.

Weed Management: Mulching may be done to retain soil moisture and suppress weed growth.

Harvesting: Plants are ready for harvest within 2-3 months after sowing seeds. Generally, terminal, succulent, lateral shoots are harvested for vegetable purposes. Harvesting may be continued for several months as and when new growths appear.

Yield: Yield may vary between 2.5 – 3 tons per hectare per harvest. In other words, 2.5 kilograms of leaves may be harvested from one square meter area during one harvest. Annual yield is approximately 25-30 t/ha per year.

Note: Detailed information on all types of spinach herbs is available in my book titled **"Spinach Herbs"**

Ginger

Scientific name of ginger is *Zingiber officinale*. Synonyms are *Amomum zingiber* and *Zingiber missionis*. It belongs to the family Zingiberaceae. Ginger is actually an herbaceous perennial crop but for commercial production it is grown as an annual herb. Ginger is tropical and subtropical in its growth habit. It is grown for its aromatic rhizomes which are used as a vegetable, a spice and as a traditional medicine. Ginger rhizomes are often called 'ginger root' though it is not actually a root. Rhizome is actually a modified underground stem of the plant.

Origin and Distribution: Ginger is believed to be originated in South East Asia, particularly in the region comprising of India and China.

Taxonomy: A detailed account of taxonomic classification of ginger plant is as given below:

Kingdom	Plantae
Class	Equisetopsida
Subclass	Magnoliidae
Superorder	Lilianae
Order	Zingiberales
Family	Zingiberaceae
Genus	Zingiber
Species	Officinale

Botanical Description: New plant grows from rhizomes planted underground. Upon maturity, each ginger plant reaches up to a height of one meter. Stem is covered by leaf sheaths. Leaves are light green in colour; and long and lanceolate in shape with a marked midrib. Each leaf is about 15 cm in length. Small yellowish flowers are borne in dense spikes. The most important plant part is edible rhizome. Fresh rhizome has a pale yellowish interior with skin colour varying from brownish dark to off-white.

The plants reach maturity within 8-10 months after planting. Harvesting time can be determined when the green leaves begin to turn yellow. Harvesting is normally done after all leaves are dried and withered away.

Economic Importance of Ginger: Ginger is used for food and drink purposes, medicinal purposes and cosmetic uses. Raw ginger is used as a vegetable and as a food flavoring agent. Ground ginger paste is used to add taste and flavor to non-vegetarian food preparations. Dry ginger has great commercial significance. Ginger powder is used in many traditional medicinal preparations, particularly in Ayurvedic medicines.

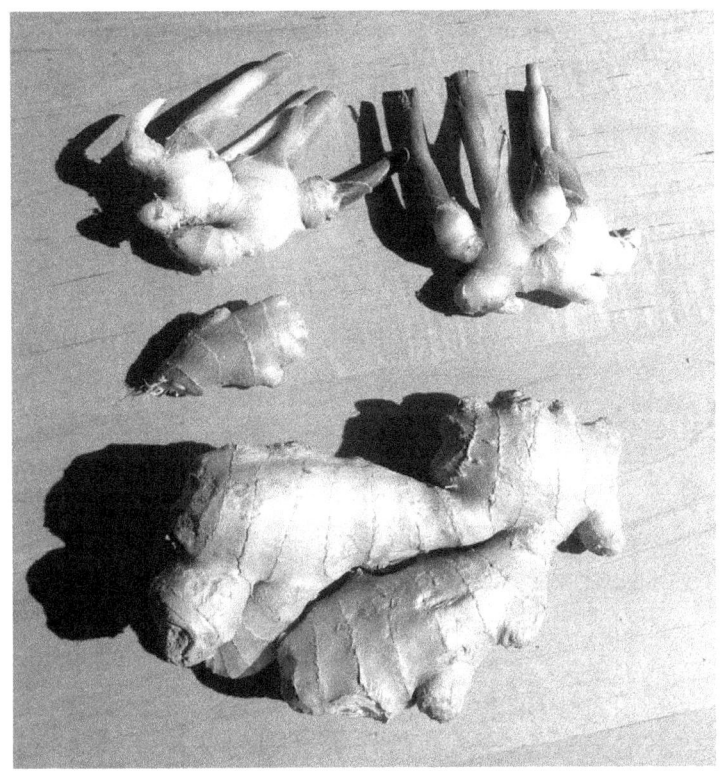

Dry Ginger: Fresh ginger rhizomes are peeled and dried in the sun to obtain commercially important 'dry ginger'. Dry ginger is marketed either as 'black' or 'white' where black dry ginger is dried ginger rhizomes with its skin on while white dry ginger is dried ginger rhizomes with its skin peeled off. Dried ginger is available in the market as whole or sliced or powdered forms. Dry ginger powder is a major ingredient in many curry powder preparations and Ayurvedic medicinal preparations. In many parts of the world, dried ginger is used to prepare ginger bread, cakes and biscuits.

Food Uses of Ginger: A detailed account of various food uses of ginger is as given below:

Raw Ginger as a Vegetable and a Fiery, Pungent and Spicy Food Flavouring Agent: Raw ginger is known as 'fresh ginger' which imparts the freshest and purest ginger flavour to food preparations. Sliced ginger is used for pickle preparations, chutneys and for making ginger curry pastes. Nutrition in 100 grams of edible portion of fresh ginger is given below:

Nutrient	Unit	Value per 100 g
Water	g	78.89
Energy	kcal	80
Protein	g	1.82
Total lipid (fat)	g	0.75
Carbohydrate, by difference	g	17.77
Fiber, total dietary	g	2
Sugars, total	g	1.7
Calcium, Ca	mg	16
Iron, Fe	mg	0.6
Magnesium, Mg	mg	43
Phosphorus, P	mg	34
Potassium, K	mg	415
Sodium, Na	mg	13
Zinc, Zn	mg	0.34
Vitamin C, total ascorbic acid	mg	5
Thiamin	mg	0.025
Riboflavin	mg	0.034
Niacin	mg	0.75
Vitamin B-6	mg	0.16
Folate, DFE	µg	11
Vitamin B-12	µg	0
Vitamin A, IU	IU	0
Vitamin E (alpha-tocopherol)	mg	0.26
Vitamin K (phylloquinone)	µg	0.1
Fatty acids, total saturated	g	0.203
Fatty acids, total monounsaturated	g	0.154
Fatty acids, total polyunsaturated	g	0.154

Ground Ginger for Culinary Purposes: Ground ginger is used as a spice in culinary preparations. It adds flavour, taste and thick gravy to many spicy food preparations particularly non-vegetarian food preparations such as meat and fish preparations. Nutrition in 100 grams of edible portion of ground ginger is given below:

Nutrient	Unit	Value per 100 g
Water	g	9.94
Energy	kcal	335
Protein	g	8.98
Total lipid (fat)	g	4.24
Carbohydrate, by difference	g	71.62
Fiber, total dietary	g	14.1
Sugars, total	g	3.39
Calcium, Ca	mg	114
Iron, Fe	mg	19.8
Magnesium, Mg	mg	214
Phosphorus, P	mg	168
Potassium, K	mg	1320
Sodium, Na	mg	27
Zinc, Zn	mg	3.64
Vitamin C, total ascorbic acid	mg	0.7
Thiamin	mg	0.046
Riboflavin	mg	0.17
Niacin	mg	9.62
Vitamin B-6	mg	0.626
Folate, DFE	µg	13
Vitamin B-12	µg	0
Vitamin A, IU	IU	30
Vitamin E (alpha-tocopherol)	mg	0
Vitamin K (phylloquinone)	µg	0.8
Fatty acids, total saturated	g	2.599
Fatty acids, total monounsaturated	g	0.479
Fatty acids, total polyunsaturated	g	0.929

Using Fresh Ginger as Ginger Preserves: Fresh, tender ginger rhizomes are cleaned and peeled before slicing them into small pieces. These small chunks of ginger are cooked in sugar syrup. After cooling the syrup, this hot and spicy ginger preserve is bottled in air tight containers and kept for further use.

Beverage Uses of Ginger: Fresh ginger is used to prepare ginger tea, ginger beer and ginger wine. Ginger tea has great medicinal benefits. It alleviates cough and cold and soothes the digestive system.

Ginger Oil (Gingerol) for Medicinal and Cosmetic Benefits:
Gingerol, an essential oil extracted from both fresh and dried ginger rhizomes have numerous medicinal and aromatic uses. Gingerol is sudorific (encourages sweating); expectorant (loosens and expels phlegm from lungs); stomachic (improves digestion and cures stomach disorders); and Emmenagogue (brings on menstruation). Ginger oil encourages sweating in high temperature fevers and cures cough and common cold.

Health Risks: Ginger paste may be applied as a topical ointment but it is not recommended because it may cause allergic reactions to some people. Ginger should be used with great precaution during pregnancy.

Growing Practices for Ginger: Major commercial varieties of ginger are 'China', 'Maran', 'Himachal', 'Rio-de-Janeiro', 'Nadia', 'Thingpuri', 'Narasapattam', 'Wynad Manantoddy', 'Karkal', 'Vengara', 'Ernad Manjeri', 'Ernad Chemad'and 'Burdwan'.

However in international spice trading markets, one can see ginger varieties named after the places of their production. In these markets, Jamaican ginger (ginger produced in Jamaica) is considered as the best variety. Then there is Kenya ginger which has also got good market. African and Indian ginger varieties which have comparatively darkish brown skin are considered inferior in quality. Another ginger variety is Japanese ginger.

Climate: Ginger is a shade-loving, tropical plant with preference for wet to moist areas; it cannot stand direct scorching sunlight, frost, very low temperatures, waterlogged soil and strong winds. Ideal climate for ginger cultivation is warm, humid, tropical and subtropical climate. It can be grown from MSL (mean sea level) up to an altitude of 1500 meters.

Soil: Well-drained, rich and moist soil with plenty of organic matter is the most ideal soil for ginger cultivation.

Light: Ginger prefers filtered sunlight and hence grows well in partially shaded locations.

Crop Rotation: Ginger can be grown both as a rainfed crop and as an irrigated crop. In irrigated crop, ginger crop can be rotated with turmeric, low-growing vegetables such as onion and garlic, fruit trees such as plantain and banana, betel vine crop, chillies, groundnut, maize and sugarcane. In rainfed crop, ginger crop may be rotated with tapioca, sweet potato, yam and dry paddy once in 3 or 4 years.

Ginger as an Intercrop: Ginger crop may be raised as an intercrop among coffee, coconut, areca nut and orange plantations.

Ginger as a Mixed Crop: Ginger crop may be raised as a mix crop along with shade-giving plants such as plantain, banana, and tree castor.

Propagation: Propagation is by planting rhizome (modified underground stem) cuttings directly in the main field. Each rhizome cutting weighing 15-30 grams and having at least 2-3 well developed growth buds is planted.

Field Preparation: Field is prepared by ploughing twice or thrice followed by levelling and ridge making. It is always better to plant ginger rhizomes on the raised beds.

In tropical countries such as India, land preparation begins during March and April. Land is ploughed into a deep fine tilth and raised beds of 15 cm height and 1 meter width and of convenient length up to 5 or 6 meter are prepared. Soil solarization is done before planting as a preventive measure.

Planting for Rainfed Crop: A spacing of 30 cm is given between two ridges/raised beds. Shallow pits are prepared on the ridges to plant rhizome cuttings which are planted in two rows. Row to row distance is 30 cm and plant to plant distance within a row is 15-20 cm.

Planting for Irrigated Crop: A spacing of 45-50 cm is given between two ridges/raised beds. Shallow pits are prepared on top of the ridges in a single row to plant rhizome cuttings. Plant to plant distance within a row is 25 -30 cm.

Sowing Depth: Rhizome cuttings are normally planted approximately 5- 10 cm deep into the pits with the growing buds up and then it is covered by a thin layer of soil.

Rate of Planting Materials Required: Approximately 2 tons of rhizome cuttings are required to plant one hectare of land. That is, about 750-1000 Kg seed rhizomes are required to plant one acre.

Planting Time: Late winter or Early Spring (March to May) is the best time for planting ginger rhizomes.

Fertilizer and Manure Application: Ginger is a heavy feeder and hence requires heavy manuring. Application of 5-10 tons of farmyard manure or compost per acre at the time of field preparation is recommended. Subsequently, vermicompost or any other organic fertilizer available locally, may be applied @500 Kg/acre. Recommended N, P, K (nitrogen, phosphorous and potassium) fertilizer doses are 75:50:50 Kg/hectare. Whole phosphorous and half potassium fertilizers are applied at the time of planting. Remaining half potassium and half nitrogen fertilizers are applied two months after planting. Remaining half dose of nitrogen fertilizer is applied one month after second application of fertilizers. In some parts of the world, farm yard manure or neem cake or castor cake @4-5 t/ha is used in 2-3 top dressings.

Irrigation: For irrigated crop, first irrigation is done soon after planting; subsequent irrigations are given just to keep the soil moist throughout the growth phase of ginger plants. Water logged or soggy soils should be avoided. For rainfed crop, moist leaf mulch (of dried or green leaves) is spread over the beds soon after planting. In certain areas, farm yard manure is used as mulch. New shoots emerge within 2-3 weeks after planting.

Mulching of the Ginger Beds: Mulching is an important cultural practice while growing ginger. At least two or three mulching are required during the growth period of a ginger plantation. First mulching is done soon after planting rhizome cuttings. Mulching with green leaves is highly recommended as this practice is proved to be more beneficial for ginger growth. Second mulching is done after first weeding and hoeing practices which is about 40-50 days after planting. Third mulching is done after second weeding and hoeing practices which is about 40-50 days after second mulching. Subsequent weeding and hoeing, and mulching practices are done as and when necessary.

Interculture and Aftercare: Hoeing and weeding are done to keep the field weed-free. It is recommended to mulch the beds thickly with biodegradable mulch soon after planting rhizome cuttings. This biomulch effectively controls weeds as well as conserves moisture.

Insect-Pest Management: Ginger is susceptible to spider mite attack in a dry weather; organic insecticides based on pyrethrum or tobacco extracts may be used to control them. Other major pests that are found attacking ginger are shoot borer (Conogethes punctiferalis/ Dicrhosis punctiferalis), nematodes, and white grub (Holotrichia setticolis). For controlling shoot borers, regular field surveillance is required which needs to be followed up by proper phyto-sanitary measures. Another option is hand picking of caterpillars and destroying them. Some growers grow neem trees in ginger plantations for its insect-repellant effect.

For controlling nematodes, application of neem cake @1ton/ha is recommended. Two applications are required one during planting and second application 45 days after planting. White grub (Holotrichia setticolis) may be controlled by tillage of fields particularly during summer and by solarization of fields. Setting up of bird perches, other bird attractants and light traps and handpicking of infested leaves and grubs may also effectively control white grub infestation.

Disease Management: Soft rot or rhizome rot caused by *Pythium aphanidermatum* and *Pythium myriotylum* is a major disease found affecting ginger rhizomes. For effective control of this disease the following cultural practices may be followed: selection of planting materials from disease free areas; ensuring proper drainage of fields; soil sterilization by solarization; sanitation of fields by burning of infected plants; removal of affected plants; application of *Trichoderma viride* at the time of planting mixed with farm yard manure; and restricted use of popular fungicide Bordeaux mixture (1%) in the areas susceptible to this disease.

Another major ginger disease is bacterial wilt caused by *Ralstonia solanacearum*/ *Pseudomonas solanacearum*. Selection of disease free planting materials and crop rotation of ginger with maize, cotton, and soybean may effectively control this disease.

Days to Maturity: It takes about 8-10 months for a ginger plantation to reach maturity.

Harvesting: Harvesting is done when the leaves have completely withered and the rhizomes are still tender and immature with outer skin still has a slight greenish colour. Rhizomes are carefully lifted by a digging-fork. Care is taken while lifting the rhizomes to avoid any sort of bruises and mechanical injuries of the rhizomes.

Curing of Freshly Harvested Ginger Rhizomes: Freshly harvested ginger rhizomes are transported to pack houses where they are cleaned. All dirt and adhering fibrous roots are removed. Rhizomes with plump buds normally called as 'seed gingers' are segregated and stored for next planting season. Remaining commercially acceptable rhizomes are graded and packed before marketing them as 'fresh ginger'.

For dry ginger, fresh gingers are soaked in water to facilitate peeling process. Peeled rhizomes are dried in the sun before marketing them as 'dry ginger'.

Yield: Yield varies depending on the variety, soil fertility, cultural practices and prevailing climatic conditions. Approximately a ginger plantation may yield 10-15 tons of fresh ginger per hectare. Dry ginger recovery is about 20-30 percent depending upon the variety used.

Storage: Fresh ginger can be stored in a refrigerator or in cold storage for several weeks without losing its freshness. Dried and powdered ginger may be stored in airtight containers in a cool, dry place for a number of years.

Seed ginger is best stored by keeping them in pits which are dug in a cool place, away from sun and rain. Before storing them, they are treated with an effective fungicide as a preventive measure against soft rot and other soil-borne fungal infections. Seed gingers thus treated are then dried in shade and placed in pits which are about one meter in depth. A layer of sand or saw dust is placed in the pit before placing seed gingers and after placing the seed gingers, the pit is covered with a wooden plank while providing maximum aeration for the stored gingers.

In higher altitudes, seed ginger may safely be stored in an underground storage chamber until next planting season.

Storage Disorders: Rhizome rot (common among bruised rhizomes), shrivelling and drying of rhizomes, dry rot and sprouting of rhizomes are observed during storage.

Grading: Grading of ginger is mainly done depending on its dry matter content and fiber content. Ginger roots having highest dry matter and lowest fiber is regarded as the best grade or Grade 1. Poorest graded ginger roots contain higher fiber content and low dry matter content.

Turmeric

Scientific name of turmeric is *Curcuma longa*. It belongs to the family Zingiberaceae, the ginger family. As in case of all plants belonging to ginger family, turmeric also prefers tropical and subtropical moist climate for its growth. Turmeric plant is an herbaceous perennial crop mainly grown for its edible rhizomes which are used as an important spice, condiment and dye. Turmeric is also known as 'Indian saffron'.

Origin and Distribution: Turmeric is believed to be a native of South East Asia, particularly India and China. It is cultivated as a commercial crop in West Bengal of India, South India, China, Taiwan, Sri Lanka, Indonesia, Peru, Jamaica, Australia and the West Indies.

Botanical Description: Turmeric is a perennial herb growing from an underground rhizome with an erect unbranched stem. Stem is covered by leaf sheaths and grows up to a height of one meter upon full maturity. Leaves are long and lanceolate in shape just like ginger leaves but much larger and broader than ginger leaves. Leaves are bright green in colour. Yellowish flowers are borne in dense spikes. The most important plant part is rhizome, which is a modified underground stem with yellow flesh. Rhizomes are ready for harvesting within 8-10 months after planting.

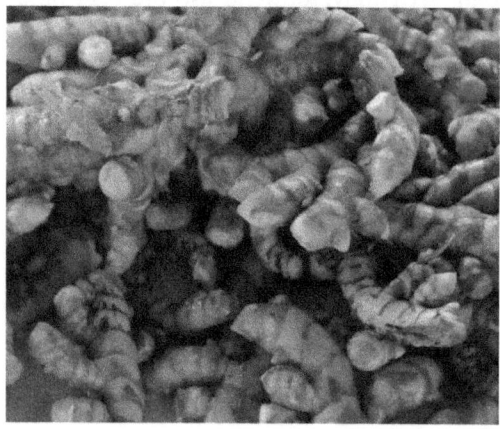

Food Uses of Turmeric: There are mainly two forms of whole turmeric available in the market: fresh turmeric and dried turmeric. Dried turmeric is used to prepare turmeric powder. Turmeric powder is a major ingredient in many curry powder preparations. In almost all Indian food preparations, turmeric powder is used as a culinary dye to impart its specific yellow colour to food. Essential oil extracted from fresh and dried rhizomes of turmeric is rich in curcumin and is used in perfume industry and food manufacturing industry.

Nutrition in Turmeric: The nutritive value of 100 grams of edible portion of turmeric powder is given below:

Nutrient	Unit	Value per 100 g
Water	g	12.85
Energy	kcal	312
Protein	g	9.68
Total lipid (fat)	g	3.25
Carbohydrate, by difference	g	67.14
Fiber, total dietary	g	22.7
Sugars, total	g	3.21
Calcium, Ca	mg	168
Iron, Fe	mg	55
Magnesium, Mg	mg	208
Phosphorus, P	mg	299
Potassium, K	mg	2080
Sodium, Na	mg	27
Zinc, Zn	mg	4.5
Vitamin C, total ascorbic acid	mg	0.7
Thiamin	mg	0.058
Riboflavin	mg	0.15
Niacin	mg	1.35
Vitamin B-6	mg	0.107
Folate, DFE	µg	20
Vitamin B-12	µg	0
Vitamin A, IU	IU	0
Vitamin E (alpha-tocopherol)	mg	4.43
Vitamin K (phylloquinone)	µg	13.4
Fatty acids, total saturated	g	1.838
Fatty acids, total monounsaturated	g	0.449
Fatty acids, total polyunsaturated	g	0.756

Medicinal Uses of Turmeric: Turmeric is anti-cancerogenic (fights against cancer-causing cells), antiseptic (kills harmful microbes), antifungal (cures fungal infections) and antiviral (cures viral infections). The active ingredient that imparts turmeric its medicinal properties is called *curcumin* which is present in both fresh and dried rhizomes of turmeric. Turmeric is also consumed internally as a stimulant.

Turmeric is anti-inflammatory and curcumin is a powerful antioxidant. Food rich in antioxidants are good as an anti-ageing diet. Antioxidants are also helpful in fighting cancerous cells. Anti-inflammatory property of curcumin is used in treating arthritis.

Turmeric is antifungal and antiseptic because of the presence of curcumin. Hence turmeric paste may be applied to cure leech bites, ring worms, skin inflammations, mouth sores, wounds, insect bites etc.

Growing Practices for Turmeric: Some of the major varieties of turmeric are 'Aleppey Turmeric', 'Madras Turmeric', 'Lokhandi', 'Duggirala', 'Tekurpeta', 'Kasturi Pasupa', 'Armoor', 'Roma', 'Suroma' and 'Chaya Pasupa'. Turmeric varieties with more than 5% curcumin content and having lemon yellow or orange yellow interior are preferred in the international markets.

There is a highly aromatic variety of turmeric available in the market which is called 'kasturi turmeric'. Its scientific name is *Curcuma aromatica.* It has great cosmetic value and used extensively in cosmetic industries for preparing face creams, facial scrubs etc.

Climate: Turmeric needs a warm and humid climate. It can be grown as a rainfed crop in heavy rainfall areas. In other areas it is grown as an irrigated crop. It can be successfully be grown from mean sea level (MSL) up to an altitude of 1200 meters.

Soil: Turmeric crop thrives well in well-drained fertile loamy soils which are rich in humus content. Turmeric plantation cannot withstand waterlogged soils and soil alkalinity.

Propagation: Propagation is mainly via rhizome cuttings and rhizome fingers (small tubers attached to mother rhizomes).

Crop Rotation: Crop rotation is recommended because turmeric is an exhaustive crop just like ginger. In wet and moist lands, turmeric crop may be rotated with paddy, sugarcane, banana and plantain once in 3-4 years. In plains turmeric can be grown in rotation with sugarcane, chilli, onion, garlic, wheat, pulses and short season vegetables.

Turmeric as an Intercrop: In tropics, turmeric may be grown as an intercrop with mango, coconut, areca nut, jackfruit tree and litchi plantations.

Field Preparation: Land is ploughed 3-4 times to bring the soil to a fine tilth. Then raised beds or ridges of one meter width and length of convenient size with a height of 15 cm are prepared with a spacing of 30-45 cm between two ridges.

Planting for Rainfed Crop: A spacing of 30-45 cm is given between two ridges/raised beds. Shallow pits are prepared on the ridges to plant rhizome cuttings which are planted in two rows. Row to row distance is 30 cm and plant to plant distance within a row is 15 cm.

Planting for Irrigated Crop: A spacing of 45-60 cm is given between two ridges/raised beds. Wider spacing is allowed to facilitate irrigation process. Shallow pits are prepared on top of the ridges in a single row to plant rhizome cuttings. Plant to plant distance within a row is 15-25 cm.

Planting Materials: Both the mother rhizomes and fingers are used as planting materials. Mother rhizomes are planted as such or split into two or more parts and used for planting. Fingers are cut into 5 cm long cuttings along with at least one healthy bud in order to be used as planting materials.

Sowing Depth: Rhizome cuttings or fingers are normally planted approximately 5- 10 cm deep into the pits prepared on the ridges with the growing buds up and then it is covered by a thin layer of soil.

Rate of Planting Materials Required for Planting: About 2-3 tons of rhizomes and fingers are required to plant one hectare of land.

Planting Time: In tropics, turmeric is planted during April to July.

Fertilizer and Manure Application: Turmeric plants require heavy manuring. Chemical fertilizers are seldom used in turmeric plantations. Turmeric is mostly grown with organic manures and biofertilizers. First application of organic manures i.e. FYM (farm yard manure) or compost is done at the time of field preparation@4-5 tons/hectare. Then soon after planting, groundnut cake or neem cake is applied @1-2 tons/hectare in two equal doses. First application is 2 months after planting and second application is done 2 months after first application. Vermicompost or coir pith compost or any other organic manure may be supplemented with the application of compost or FYM.

Irrigation: Irrigation is done in the furrows available between two ridges/raised beds. For irrigated crop, first irrigation is done soon after planting; subsequent irrigations are given just to keep the soil moist throughout the growth phase of the plants. Water logged or soggy soils should be avoided. For rainfed crop, moist leaf mulch (of dried or green leaves) is spread over the beds soon after planting. In certain areas, farm yard manure is used as mulch. New shoots emerge within 2-3 weeks after planting.

Mulching of the Turmeric Beds: Mulching is an important cultural practice while growing turmeric. At least two or three mulching are required during the growth period of a turmeric plantation. First mulching is done soon after planting rhizome cuttings. Mulching with green leaves is highly recommended as this practice is proved to be more beneficial for turmeric growth. Second mulching is done after first weeding and hoeing practices which is about 40-50 days after planting. Third mulching is done after second weeding and hoeing practices which is about 40-50 days after second mulching. Subsequent weeding and hoeing, and mulching practices are done as and when necessary.

Interculture and Aftercare: Hoeing and weeding are done to keep the field weed-free. It is recommended to mulch the beds thickly with biodegradable mulch soon after planting rhizome cuttings. This biomulch effectively controls weeds as well as conserves moisture.

Insect-Pest Management: Shoot borer is a major insect found attacking turmeric plantations. In order to control shoot borers the infected shoots may be removed and destroyed. Neem oil 5% spray once every 2 weeks is found to be effective in controlling shoot borers.

Disease Management: Leaf spot and leaf blotch are two major diseases of turmeric. These fungal infections may be treated successfully by the restricted application of fungicide Bordeaux mixture (1%). Another major disease is rhizome rot. It can be controlled either by soil solarization or by *Trichoderma* application to soil at time of planting.

Harvesting: Turmeric crop will be ready for harvesting within 8-10 months after planting. Turmeric is harvested when green leaves have completely turned yellow and started withering. Dried leaves are cut close to the ground. Then land is irrigated if necessary to facilitate harvesting process. The rhizomes are dug and taken out by using a digging fork or crowbar or a spade. Alternatively, the rhizomes may be hand-picked after ploughing the land.

Curing Turmeric: Freshly harvested turmeric rhizomes are transported into pack houses where they are cleaned from dirt and all attached fibrous roots are removed. Fingers are separated from mother rhizomes and kept for curing before selling in the markets. Mother rhizomes are normally used as planting materials for next season planting.

Finger turmeric is boiled in drums until white fumes appear giving out characteristic earthy turmeric odour. Then the cooked fingers are spread on a clean floor or mat in the sun in order to dry them. Generally two weeks of sun drying is required to obtain properly cured turmeric fingers. The market of turmeric depends on its curcumin content and colour. Proper curing process is essential in order to ensure proper colour development and quality in the cured turmeric.

Storage: Fresh turmeric may be stored in a refrigerator or in cold storage for several weeks. Turmeric powder if stored in an air tight container in a cool place will last for several years. Cured dried turmeric can be stored in a dry cool place in gunny bags.

Yield: Approximately 20-25 tons of turmeric is obtained from one hectare turmeric plantation. Cured turmeric is about 25 percent of the fresh rhizomes by weight.

Note: More detailed information on ginger and turmeric is available in my book titled **"Ginger, Turmeric, and Indian Arrowroot"**

Rhubarb

Rhubarb is popularly known as 'the pie-plant' because of its significance as a major ingredient in preparing pies. Botanical name of Rhubarb is *Rheum rhabarbarum*. Rhubarb is a cool season crop which is grown as an herbaceous, perennial vegetable. Rhubarb belongs to the family Polygonaceae. The economically significant portion of the plant is its fleshy leafstalks and the plant is normally grown for these fleshy leaf stalks (petioles) only. Colour of petioles may be red, white or green depending upon the cultivar.

While the leafstalks are edible, its thick large triangular-shaped leaves are not edible because the leaves contain large amounts of oxalic acid which is toxic for human consumption. Hence leaves should not be eaten. Rhubarb is a short plant with its leaves grow from thick, fleshy rhizomes. Rhizomes are modified underground stems.

Food Uses: Rhubarb leaf stalks are used for preparing desserts, jams, stews, sauces and pies in many parts of the world. Rhubarb leaf stalks are normally cooked with sugar because of its strong, tart flavour. Frost-damaged rhubarb leafstalks are not recommended for food uses as they may contain high amounts of oxalic acid.

Medicinal Uses: In traditional Chinese medicines rhubarb is used as a laxative. It can also be used as a dieting aid.

Origin: Rhubarb is believed to be originated in cooler parts of Asian continent.

Types of Rhubarb: There are two types of rhubarbs, green-stalked and red-stalked. Red or pink stalked varieties are rich in anthocyanins and are more popular among consumers. However, green-stalked varieties are more robust and they yield more. Popular red –stalked varieties are Canada Red, Cherry Red, Crimson Red, MacDonald, Ruby, Mammoth Red and Valentine. Victoria is a major green-stalked variety popular in the United States of America.

Growing Practices: Rhubarb is a hardy, frost-resistant plant. It can be grown either as a field crop or as a forced crop in cooler forcing sheds. Rhubarb grows best in cooler climatic conditions. It is resistant to cold and dry weather conditions (drought). Favorite temperature range for rhubarb growing is below 40°F in the winter to break dormancy and stimulate spring growth; and below 75°F in the summer for vigorous vegetative growth. Extreme weather conditions such as hot dry weathers and extreme colder climates are not suitable for successful rhubarb cultivation.

Soil Requirements: Rhubarb grows well in well-drained, fertile soils that are rich in organic matter. Slightly to moderately acidic soils are also good for rhubarb cultivation as rhubarb plants are tolerant to soil acidity. Ideal pH is 6.0 to 6.8.

Propagation: Rhubarb is propagated by crowns. Divisions of rhizomes with healthy growing buds are called 'crowns'. Normally, a division of rhizome containing one or two healthy growing buds is selected for planting. One-year old crowns are recommended.

Propagation by seeds is also possible but it is not recommended as seedlings thus produced may not be true-to-type. Another disadvantage of seed propagation is, seedling-propagated plants produce low-quality, and small-sized leaf stalks.

Selection of Crowns

1. Only vigorous, healthy, true to type crowns should be selected
2. Older crowns and any off types should be discarded
3. Avoid diseased and damaged crowns

Site Preparation: Site where full sunlight is available is best suitable for rhubarb growing. Site should be well prepared by ploughing and levelling. All types of weeds, both annual and perennial, should be removed as rhubarb is a perennial crop which remains on the site for several years. As a general rule, a clean site where there is no threat of perennial weed infestation is preferred for rhubarb cultivation.

Planting Time: Rhubarb crowns are planted either in early spring when the roots are still dormant or in the fall after dormancy has set in.

Planting Method: While planting, crown buds are placed minimum 2 inches below the soil surface. Planting on ridges or raised beds ensures good water drainage which in turn prevents occurrence of crown rot. Soon after planting the crowns on beds, they are covered with soil and a light irrigation is given thereafter.

Spacing: Recommended spacing is 0.5 to 1 meter within rows and 1 to 1.5 meter between rows.

Fertilizer Application: Rhubarb plants are heavy feeders and hence liberal amounts of fertilizers must be incorporated into the soil. Fertilizer application should be based on soil test analysis (which shows nutrient reserves present in the soil) as well as based on crop requirements.

At the time of site preparation, a good amount of organic manures (15-20 tonnes/acre) should be incorporated into the top soil. Thereafter 50 – 60 Kg N (nitrogen), 25 Kg P (phosphorus), and 50 Kg K (potassium) should be added annually in two split doses: first dose during active vegetative growth and second dose after harvesting.

Irrigation: Rhubarb is a moisture loving plant. Organic bio mulches are recommended for soil moisture conservation. During dry periods, frequent irrigation is necessary. Irrigation should be done based on soil moisture level conditions and prevalent climatic conditions. Trickle irrigation may help conserve water and may also allow fertigation (supplying nutrients along with irrigational water).

Weed Control: Perennial weeds may be a problem while growing a rhubarb plantation. Growers may adopt mechanical and cultural weed control methods such as mulching to control the weeds. When crowns start growing and leaves start appearing, mulching is recommended to preserve the soil moisture as well as to control the weed growth.

After care: Since rhubarb is a perennial crop, the plantation may get crowded over a period of time. So division of individual plants is recommended once in every three to four years. This is done during winter months while plants are dormant.

During dividing individual plants care is taken not to damage the growing buds. Normally, each plant is divided into three or four separate plants. New plants may be used to plant a new plantation. Flower stalks or seed stalks should be removed as and when they appear.

Economic Life of a Rhubarb Plantation: Good yield is obtained from third year onwards. Under good cultural management practices, a rhubarb plantation may last up to 12 to15 years.

Harvesting: Harvesting is done from the third year of planting onwards. During harvest, leafstalks of 12 to 18 inches long are harvested. Each harvest period may last up to 8 to 10 weeks for a large plantation. During harvesting process leafstalks (petioles) of rhubarb are carefully pulled out without damaging the leafstalks and developing buds. While harvesting a few leafstalks are left intentionally on the plant to ensure further growth of the plant. Never pull out the leafstalks heavily as this may weaken the crowns from the base.

Immediately after pulling out the leaf stalks, leaf blades are separated leaving 3 to 4 cm of midribs with intervening tissues (leaf lamina). Some leaves with stalks should be kept in the plant. That is, individual plants should never be stripped of entire leaves while harvesting. This will ensure new growth. New stalks come up within a few weeks (within 7 to 10 days).

Harvesting and Yield: Normally rhubarb is harvested from mid-May through August. Estimated yield from a well-managed rhubarb plantation is 15 to 18 tonnes per acre.

Diseases: A detailed account of various diseases affecting rhubarb plant is as given below:

Disease	Symptoms	Control
Botrytis rot (Botrytis cinerea)	Common in forced rhubarb; upper part of the leaf stalks are affected; upper part start rotting because of the extreme moist conditions (high temperature with high humidity)	✓ Good ventilation in forcing sheds ✓ Careful handling of crowns while lifting and placing them in forcing sheds to avoid damage and injury to crowns ✓ Removal of weak and damaged shoots
Crown rot (Erwinia rhapontici)	Crown rot is mainly due to the presence of over-wet soil around the plants. Symptoms appear at the base of the leafstalks and crowns slowly turn brown and later rotting begins	✓ Control measures for crown rot include destroying infected plants, providing good water drainage, field sanitation, and treating roots of the crowns with formaldehyde solution before planting
Downy Mildew (Peronospora jaapiana)	It causes serious leaf damage and affect petiole quality	✓ Avoid infected sites for planting crowns ✓ Since this is a soil borne disease, the best option is to remove and burn infected crowns
Leaf Spot	This is another fungal infection found affecting the leaves of the rhubarb plants	✓ Leaf spot disease can be controlled by the application of Bordeaux mixture in spring when vegetative growth of the plant begins

Insect Pests: A detailed account of various insects and pests affecting rhubarb plant is as given below:

Insect Pests	Symptoms of Attack	Control
Rhubarb beetle (*Rhubarb curculio*), a snout beetle	They bore into the leaf stalks, crowns and roots of the plants	Control measure includes handpicking the insects and destroying them; burning all badly affected plant parts and plants; and using recommended organic insecticides
Rosy Rustic Moth (Hydraecia micacea)	They form tunnels at the base of leafstalks	Field monitoring and destroying the weed plants where the moth lays eggs
Slugs and Snails (Derocerus, Milax, Helix and others)	They eat into the leaf stalks	✓ Avoid poorly drained soils as damp moist soil attracts slugs and snails ✓ Avoid planting crowns that are infested with slugs and snails
Stem nematode (Ditylenchus dipsaci)	Base of infested leaf stalks gets swollen and then split	✓ Avoid fields infested with nematodes for planting crowns ✓ Field hygiene and good water drainage are recommended. ✓ Use only clean, nematode-free crowns for planting ✓ Destroy infected crowns by burning

Precooling: Pulled leaf stalks should immediately be transported to pack houses and be kept in cool shady place for some time to remove field heat. Then the produce is precooled to 0 °C (32 °F) either by hydrocooling or by forced air cooling.

Optimum Storage Conditions: Freshly harvested rhubarb stalks are best stored at 0° C (32 °F) and at 95 to 100 percent relative humidity. Under these optimum conditions, rhubarb leaf stalks can be stored for 2 to 4 weeks.

Packaging for Storage: Freshly harvested leafstalks are cleaned (never wash the fresh rhubarb petioles), trimmed, tied into bundles and packed in either perforated polyethylene bags or plastic crates before delivering the produce at processing centres where the produce is kept refrigerated at optimum storage conditions. Maximum shelf life under these conditions is up to 4 weeks.

Grading: Grading of field grown rhubarb leafstalks is done on the basis of leafstalk appearance, color of the leafstalks and the presence of produce damages and defects. A list of major U.S. grades for field-grown rhubarb is given below:

Grade	Description
U.S. Fancy	The best quality produce with glossy bright colored leafstalks and free from all damages and defects
U.S. No. 1	The second best quality after U.S. Fancy
U.S. No. 2	The grade after U.S. Fancy and U.S. 1
Unclassified	The produce which cannot be grouped into any of the above categories

Packaging for Markets: For retail and wholesale markets, rhubarb petioles are packed in 4.5 Kg, 6.8 Kg, or 9.0 kg cartons.

Quality Indices: Good quality rhubarb leaf stalks are fresh in appearance without any decay or desiccation, bright colored (red, white or green depending upon the cultivar), glossy, crisp and firm.

Postharvest Diseases: Post harvest diseases are a major reason for large quantities of post harvest losses of the crop. Most of the post harvest diseases are due to poor sanitation of the storage units and by the use of poor quality water. So the best control measure for these diseases is to ensure that proper sanitation is maintained in storage units and the produce is kept at optimum storage conditions. A list of major post harvest diseases of rhubarb is given below:

Disease	Symptoms
Anthracnose (Colletotrichum erumpens)	Causes oval, soft, watery lesions on petioles
Bacterial soft rot (Pseudomonas marginalis, Erwinia caratovra)	Causes a soft, slimy decay
Gray Mold (Botrytis cinerea)	Causes soft, brown lesions on petioles

Physiological Disorders: A list of major post harvest disorders of rhubarb is given below:

Disorder	Description
Splitting of leafstalks	Petioles that are completely devoid of leaf lamina are susceptible to splitting when exposed to moisture during storage
Pithiness of Petioles	Over-mature petioles become pithy during storage
Abrasion of Petioles	Abrasion of petioles during field handling adversely affects appearance

Greenhouse Cultivation: Rhubarb is suitable for greenhouse cultivation and the crop that is grown in heated greenhouses is known as hothouse rhubarb. Greenhouse rhubarbs are sweeter, tender and brighter colored than that of field-grown rhubarbs.

Forcing Crowns During Off Season: During off season, crowns may be lifted from the ground and placed in forcing shed in order to force the production of good quality leafstalks. Forcing is done by exposing crowns to low temperatures (below 5 Degree Celsius) in cold units during colder months of October to December. For forcing, use only vigorous, healthy, damage-free and disease-free crowns. It is believed that if flower stalks or seed stalks are removed from the crowns before forcing crowns, higher yields may be obtained. Weekly irrigation is recommended for crowns placed in forcing sheds. But care needs to be taken not to irrigate the crop excessively as it increases humidity within the sheds which in turn may result in increased incidences of diseases. Also special care is to be taken to keep the forcing shed clean always to prevent disease-pest incidences.

Forced crop produce is available from December to March. In certain cases, growth regulator Gibberellin is recommended as a replacement for cold units.

How to Apply Gibberellins? Crowns are lifted from the ground and placed in forcing sheds. Required quantities of gibberellins are then applied on crowns during the months of November - December.

Nutrition in Rhubarb Leafstalks: Rhubarb leaves are rich in vitamins and minerals. Vitamins present in the leaves include vitamin C, vitamin A, vitamin K, vitamin B and vitamin E. Minerals present include calcium, magnesium, iron, phosphorous, potassium, sodium and zinc. A detailed account of nutrition in raw (fresh) rhubarb leaf stalks is given below:

Nutrient	Unit	Value /100g	1.0"cup, diced"122.0g	1.0"stalk "51.0g
Water	g	93.61	114.2	47.74
Energy	kcal	21	26	11
Protein	g	0.9	1.1	0.46
Total lipid (fat)	g	0.2	0.24	0.1
Carbohydrate	g	4.54	5.54	2.32
Fiber, total dietary	g	1.8	2.2	0.9
Sugars, total	g	1.1	1.34	0.56
Calcium, Ca	mg	86	105	44
Iron, Fe	mg	0.22	0.27	0.11
Magnesium, Mg	mg	12	15	6
Phosphorus, P	mg	14	17	7
Potassium, K	mg	288	351	147
Sodium, Na	mg	4	5	2
Zinc, Zn	mg	0.1	0.12	0.05
Vitamin C	mg	8	9.8	4.1
Thiamin	mg	0.02	0.024	0.01
Riboflavin	mg	0.03	0.037	0.015
Niacin	mg	0.3	0.366	0.153
Vitamin B-6	mg	0.024	0.029	0.012
Folate, DFE	µg	7	9	4
Vitamin B-12	µg	0	0	0
Vitamin A, IU	IU	102	124	52
Vitamin E	mg	0.27	0.33	0.14
Vitamin K	µg	29.3	35.7	14.9
Fatty acids, total saturated	g	0.053	0.065	0.027
Fatty acids, total monounsaturated	g	0.039	0.048	0.02
Fatty acids, total polyunsaturated	g	0.099	0.121	0.05

A detailed account of nutrition in frozen, cooked (with sugar) rhubarb leaf stalks is given below:

Nutrient	Unit	Value/100g	1.0"cup"240.0 g
Water	g	67.79	162.7
Energy	kcal	116	278
Protein	g	0.39	0.94
Total lipid (fat)	g	0.05	0.12
Carbohydrate	g	31.2	74.88
Fiber, total dietary	g	2	4.8
Sugars, total	g	28.7	68.88
Calcium, Ca	mg	145	348
Iron, Fe	mg	0.21	0.5
Magnesium, Mg	mg	12	29
Phosphorus, P	mg	8	19
Potassium, K	mg	96	230
Sodium, Na	mg	1	2
Zinc, Zn	mg	0.08	0.19
Vitamin C	mg	3.3	7.9
Thiamin	mg	0.018	0.043
Riboflavin	mg	0.023	0.055
Niacin	mg	0.2	0.48
Vitamin B-6	mg	0.02	0.048
Folate, DFE	Âµg	5	12
Vitamin B-12	Âµg	0	0
Vitamin A, IU	IU	73	175
Vitamin E	mg	0.19	0.46
Vitamin K	Âµg	21.1	50.6
Fatty acids, total saturated	g	0.014	0.034
Fatty acids, total monounsaturated	g	0.01	0.024
Fatty acids, total polyunsaturated	g	0.025	0.06

A detailed account of nutrition in frozen, UNCOOKED rhubarb leaf stalks is given below:

Nutrient	Unit	Value/100g	1.0"cup, diced"137.0g
Water	g	93.51	128.11
Energy	kcal	21	29
Protein	g	0.55	0.75
Total lipid (fat)	g	0.11	0.15
Carbohydrate	g	5.1	6.99
Fiber, total dietary	g	1.8	2.5
Sugars, total	g	1.1	1.51
Calcium, Ca	mg	194	266
Iron, Fe	mg	0.29	0.4
Magnesium, Mg	mg	18	25
Phosphorus, P	mg	12	16
Potassium, K	mg	108	148
Sodium, Na	mg	2	3
Zinc, Zn	mg	0.1	0.14
Vitamin C	mg	4.8	6.6
Thiamin	mg	0.031	0.042
Riboflavin	mg	0.029	0.04
Niacin	mg	0.203	0.278
Vitamin B-6	mg	0.025	0.034
Folate, DFE	µg	8	11
Vitamin B-12	µg	0	0
Vitamin A, IU	IU	107	147
Vitamin E	mg	0.27	0.37
Vitamin K	µg	29.3	40.1
Fatty acids, total saturated	g	0.029	0.04
Fatty acids, total monounsaturated	g	0.021	0.029
Fatty acids, total polyunsaturated	g	0.054	0.074

Parsley

Parsley (*Petroselinum crispum* formerly known as *Petroselinum hortense*) is a biennial herbal spice crop belongs to the family Apiaceae (Umbelliferae). It can successfully be cultivated both in tropical and temperate climates. Commercially it is grown as an annual and its economically significant part is its aromatic leaves. Parsley is believed to be originated in the region comprising of Europe and western Asia.

Botanical Description: A detailed botanical description of parsley plant is as given below:

Leaves	Bright green leaves; Leaf shape is triangular, varies from 3-leaflet (tripinnate) to greatly curled and divided; 10–25 cm long
Root	Tap root system
Flower	Flowering stem grows up to one meter tall that bears flat-topped compound umbels with small yellow to yellowish-green flowers

Fruits	2-3 mm long, crescent shaped, rigid
Seeds	Ovoid, 2–3 mm long, brown colored, smooth, ribbed and ovate
Plant Height at Maturity	1 to 1 1/2 feet

Types of Parsley: Two types of parsley are popular among growers, Curly-leaf parsley and Italian parsley. Curly leaf parsley is *Petroselinum crispum var. crispum*, and Italian parsley is *Petroselinum crispum var. neapolitanum*. Curly-leaf parsley is characterized by curled, crisped leaves and Italian parsley is characterized by flat, noncrisped leaves. Italian parsley is more fragrant and less bitter than the curly-leaf parsley. Both types are grown for its aromatic leaves. Chopped fresh leaves are used as a garnish in soups, salads, stews, sauces etc. Leaf flakes or dried leaves are used as a flavouring agent in many continental food preparations.

Cultivars: Mainly two types of parsley are popular among growers.

Curly-leaf Types	Banquet, Dark Moss Colored, Decorator, Deep Green, Forest Green, Improved Market Gardener, Moss Curled, Sherwood
Flat-leaf Italian Types	Plain, Plain Italian Dark Green

Economic Significance: Parsley leaves are used for culinary and medicinal purposes. Both leaves and seeds are used for essential oil extraction.

Culinary Uses: A detailed account of various culinary uses of parsley is as given below:

Fresh Leaves and Sprigs as Seasoning Agent and Condiment: Fresh parsley leaves and sprigs are used as a garnishing and seasoning agent in several food preparations such as sauces, stews, salads, vegetables and meat and fish preparations. A detailed account of nutrition in fresh parsley leaves and sprigs is given in the table below:

Nutrient	Unit	Value per100g	1.0"cup choppe d"60.0g	1.0"tbs p"3.8g	10.0"spri gs"10.0g
Water	g	87.71	52.63	3.33	8.77
Energy	kcal	36	22	1	4
Protein	g	2.97	1.78	0.11	0.3
Total lipid (fat)	g	0.79	0.47	0.03	0.08
Carbohydrate	g	6.33	3.8	0.24	0.63
Fiber, total dietary	g	3.3	2	0.1	0.3
Sugars, total	g	0.85	0.51	0.03	0.08
Calcium, Ca	mg	138	83	5	14
Iron, Fe	mg	6.2	3.72	0.24	0.62
Magnesium, Mg	mg	50	30	2	5
Phosphorus, P	mg	58	35	2	6
Potassium, K	mg	554	332	21	55
Sodium, Na	mg	56	34	2	6
Zinc, Zn	mg	1.07	0.64	0.04	0.11
Vitamin C	mg	133	79.8	5.1	13.3
Thiamin	mg	0.086	0.052	0.003	0.009
Riboflavin	mg	0.098	0.059	0.004	0.01
Niacin	mg	1.313	0.788	0.05	0.131
Vitamin B-6	mg	0.09	0.054	0.003	0.009
Folate, DFE	µg	152	91	6	15
Vitamin B-12	µg	0	0	0	0
Vitamin A, IU	IU	8424	5054	320	842
Vitamin E	mg	0.75	0.45	0.03	0.08
Vitamin K	µg	1640	984	62.3	164
Fatty acids, total saturated	g	0.132	0.079	0.005	0.013
Fatty acids, total monounsaturated	g	0.295	0.177	0.011	0.03
Fatty acids, total polyunsaturated	g	0.124	0.074	0.005	0.012

Fresh and chopped parsley leaves are rich in Vitamins, particularly vitamin K, vitamin A, folate (folic acid), and vitamin C (ascorbic acid). They are also rich in minerals particularly calcium, and potassium.

Dried Leaf Flakes as Food Flavouring Agent and Condiment

Processing parsley leaves by drying (dehydration) is recommended for extending the shelf life of the produce. Dehydrated parsley leaves or leaf flakes have a good market also.

Nutrition in Dehydrated or Dried Parsley

Nutrient	Unit	Value per100g	1.0"tsp"0.5g	1.0"tbsp"1.6g
Water	g	5.89	0.03	0.09
Energy	kcal	292	1	5
Protein	g	26.63	0.13	0.43
Total lipid (fat)	g	5.48	0.03	0.09
Carbohydrate	g	50.64	0.25	0.81
Fiber, total dietary	g	26.7	0.1	0.4
Sugars, total	g	7.27	0.04	0.12
Calcium, Ca	mg	1140	6	18
Iron, Fe	mg	22.04	0.11	0.35
Magnesium, Mg	mg	400	2	6
Phosphorus, P	mg	436	2	7
Potassium, K	mg	2683	13	43
Sodium, Na	mg	452	2	7
Zinc, Zn	mg	5.44	0.03	0.09
Vitamins				
Vitamin C	mg	125	0.6	2
Thiamin	mg	0.196	0.001	0.003
Riboflavin	mg	2.383	0.012	0.038
Niacin	mg	9.943	0.05	0.159
Vitamin B-6	mg	0.9	0.005	0.014
Folate, DFE	µg	180	1	3
Vitamin B-12	µg	0	0	0
Vitamin A, IU	IU	1939	10	31
Vitamin E	mg	8.96	0.04	0.14
Vitamin K	µg	1359.5	6.8	21.8
Fatty acids, total saturated	g	1.378	0.007	0.022
Fatty acids, total monounsaturated	g	0.761	0.004	0.012
Fatty acids, total polyunsaturated	g	3.124	0.016	0.05
Fatty acids, total trans	g	0	0	0

Parsley leaf flakes are rich in Vitamins, particularly vitamin K, vitamin A, folate (folic acid), and vitamin C (ascorbic acid). Mineral content in dried leaves is much higher than that of fresh leaves. They are rich in minerals such as calcium, magnesium, phosphorous and potassium.

Freeze-Dried Parsley Leaves as a Food Flavouring Agent

Freeze-dried parsley leaves also have a good market as a seasoning agent. Freeze-drying also helps extend the shelf life of the fresh produce.

Nutrition in Freeze-Dried Parsley Leaves

Nutrient	Unit	Value per100g	1.0"tbsp"0.4 g	0.25"cup"1.4 g
Water	g	2	0.01	0.03
Energy	kcal	271	1	4
Protein	g	31.3	0.13	0.44
Total lipid (fat)	g	5.2	0.02	0.07
Carbohydrate	g	42.38	0.17	0.59
Fiber, total dietary	g	32.7	0.1	0.5
Calcium, Ca	mg	176	1	2
Iron, Fe	mg	53.9	0.22	0.75
Magnesium, Mg	mg	372	1	5
Phosphorus, P	mg	548	2	8
Potassium, K	mg	6300	25	88
Sodium, Na	mg	391	2	5
Zinc, Zn	mg	6.11	0.02	0.09
Vitamin C	mg	149	0.6	2.1
Thiamin	mg	1.04	0.004	0.015
Riboflavin	mg	2.26	0.009	0.032
Niacin	mg	10.4	0.042	0.146
Vitamin B-6	mg	1.375	0.006	0.019
Folate, DFE	µg	194	1	3
Vitamin B-12	µg	0	0	0
Vitamin A, IU	IU	63240	253	885

Parsley Leaf Oil as a Food Flavouring Agent: Dried leaf flakes of parsley are used to extract parsley leaf oil which is rich in many economically important volatile compounds. Leaf oil is extracted through steam distillation of leaf flakes. Parsley leaf oil is a wonderful food flavouring agent.

Parsley Seed Oil as a Food Flavouring Agent: Parsley seeds contain both fixed and volatile oils. Volatile oils are more popular as a food flavouring and seasoning agent. Seed oil can be extracted from seeds through steam distillation process. Seed oil contains volatile compounds such as apiol, myristicin, tetramethoxybenzene and other compounds.

Parsley Seed Oil as a Fragrance in Cosmetics, Perfumes, Soaps and Creams: Parsley seed oil may be used as a fragrance in many cosmetic preparations, perfumes, soaps, and creams.

Medicinal Uses: Parsley fresh leaves are sued to prepare fresh juices and herbal tea which may be used as alternative medicines in certain cases of ailments. Parsley leaf (herb) oil and seed oil also have medicinal properties and in some cases used as natural medicines. List of medicinal properties of parsley is given below. This list is given for informational purposes only. Medicinal use of parsley should always be according to the advice of an expert medical practitioner in the field of alternative or natural medicines.

Parsley as a Diuretic: Since parsley is rich in potassium, it has a diuretic effect on the body.

Parsley for a strong immune system: Since parsley is rich in vitamins, it imparts a strong immune system upon regular consumption.

Parsley oil as an anti cancer agent: Since parsley is rich in volatile oil components such as myristicin, limonene, eugenol, and alpha-thujene and flavonoid antioxidants such as apiin, apigenin, crisoeriol, and luteolin, it has strong anti-cancer properties. Myristicin is believed to be having the properties of inhibiting tumor formation in the lungs. Apigenin is believed to be having strong anticancer properties.

Parsley for a Healthy Heart: Since parsley is rich in folate (folic acid), its regular consumption increases overall health of the heart. Folate or folacin reduces the level of homocysteine, a naturally occurring harmful amino acid in the blood stream. High levels of homocysteine in blood may damage blood vessels making the human body susceptible to a range of heart disorders such as stroke and heart attack. Folate present in parsley converts homocysteine in to harmless molecules.

Parsley as an Anti-Inflammatory Agent: Presence of eugenol in parsley oil imparts its powerful anti-inflammatory properties. Hence parsley oil may be used as a protection against osteoarthritis and rheumatoid arthritis.

Parsley as a Detoxifying Agent: Apigenin and myristicin, two important antioxidants found in parsley enhances functioning of human liver enzymes which, in turn, results in detoxification of human body.

Health Risks Associated With Parsley Consumption

Parsley is high (1.70% by mass) in oxalic acid. Oxalates are a group of naturally-occurring compounds in plants. Upon consumption, oxalic acid within the human body may result in the formation of kidney and gall bladder stone. Hence for people suffering from kidney and bladder stones and infections, consumption of large amounts of oxalates may pose health threats. Hence parsley is not recommended for such persons. Parsley is also not recommended for consumption by pregnant women. Parsley as an oil, root, leaf, or seed could lead to uterine stimulation and preterm labor.

Ornamental Uses: Parsley is prized for its attractive ornamental foliage. The plant is sometimes grown as an ornamental edging plant.

Growing Practices for Parsley: Parsley is short duration crop and it may take up to 70 to 90 days from seeding to first harvest of fresh leaves.

Climatic Requirements: It is a cool season, hardy crop which flourishes well both in temperate and tropical climates. It can also be produced under glasshouse or polyhouse production practices. Parsley loves full sun. It grows best between 22 °C and 30 °C.

Soil Requirements: Parsley grows well in rich, fertile, moist soil with good drainage. Ideal soil pH is between 5 and 8.

Propagation: Propagation is mainly through seeds.

Seed Rate: About 50 to 100 gm of seed is sufficient for one acre. Seed rate per acre should be determined depending on prevalent soil and environmental conditions.

Pre-Sowing Seed Treatment: Parsley seeds have very low germination rate. Seeds may take up to four to six weeks to germinate. Therefore a pre-treatment soaking of parsley seeds is recommended to hasten germination. Seeds may be soaked in hot water for 24 hours before sowing them.

Sowing Time: Best time for sowing parsley seeds is spring. Spring-sown crop is harvested in late spring through summer. Parsley may be sown during summers for fall, and winter harvesting and during falls for an early spring harvesting.

Sowing Process: Seeds may be sown in well prepared nursery beds for field transplanting or they may be directly seeded. In case of nursery bed sowing, a fine seedbed is required.

Nursery Bed Preparation: Soils need to be prepared by ploughing and levelling and incorporating organic manures such as compost and leaf mould before preparing nursery beds. Raised nursery beds are prepared at a height of at least 50 to 60 inches. At least 3 to 4 rows may be prepared per bed. A space of at least 20 inches may be left between two rows. Seeds are sown in these rows and then covered with leaf mould and sand. A light watering is done immediately after sowing.

Field Preparation: Field is prepared by ploughing, harrowing and levelling. Organic manures and fertilizers such as farm yard manure, compost and leaf mould are incorporated into the top soil to enhance soil fertility.

Transplanting: Two-month-old seedlings are used for transplanting.

Spacing: Standard spacing is 50 cm × 50 cm. Spacing may be adjusted according to the grower requirement. It is believed that high plant density yields higher economic yields.

Manuring and Fertilization: Fertilizer application needs to be based on soil testing results and crop requirements. As a general rule, application of 15 tonnes of farmyard manure, 65 kg Nitrogen, 40 kg of Phosphorous and 25 kg of Potassium per hectare of area gives a good crop yield.

Irrigation: Overhead sprinklers or drip irrigation may be used for large parsley plantations. Irrigational frequency needs to be based on prevalent climatic and soil conditions. As a general rule, crop may be irrigated at 15–20 days intervals.

Insects Pests: Major insect pests that are found affecting parsley crop includes aphids, cabbage looper or beet armyworm, carrot weevil, corn earworm, flea beetles, leafhoppers, plant bugs, and rootknot nematodes. IPM (integrated pest management) practices including mechanical traps, field monitoring, and biological pest control methods are recommended for effective control of these insect-pests. Nematodes can be effectively controlled by soil fumigation process. Soil needs to be fumigated before planting the crop in the main field.

Diseases: Major diseases that are found affecting parsley crop are leaf spot, aster yellow and damping-off. A description of these diseases is given below:

Disease	Description
Septoria leaf spot	This is a seed-borne disease caused by fungus *Septoria apiicola* Control measure: Purchase good quality infection-free seeds
Aster Yellow	This is a viral disease Control measure: Eliminate leafhoppers (disease-spreading vector) from the field
Damping-off	This is caused by fungus Pythium spp. Control measure: Since this pathogen is soil-borne, best control measure is soil treatment with a recommended fungicide

Weed Control: In organic production practices, manual and mechanical weed control practices are recommended. Hoeing may be used as a weed control method. Hoeing and weeding are done as and when required to keep the crop weed-free.

Interculture: Shallow and clean cultivation is recommended for parsley.

Harvesting and Yield: Harvesting is done either by hand or by a harvesting machine. In hand harvesting, outer leaves of marketable size are cut 2.5 to 3 centimeters above the crown by a harvesting knife. In machine harvesting, leaves are mechanically clipped 2.5 to 7.5 centimeters above the crown. Multiple harvestings are possible. Average yield per acre is about 700 -1000 bushels.

Field Packing: A bunch of harvested fresh leaves are bunched together and tied with a rubber and or a jute thread or any other similar materials and placed in crates or wooden baskets before transporting the produce to pack houses or nearest markets or processing centres.

Postharvest Management: A detailed account of various post harvest practices required for parsley is given below:

Precooling: Precooling of freshly harvested parsley leaves is recommended in order to remove the excess field heat present in the fresh produce. Normally precooling is done by hydrocooling.

Optimum Storage Conditions: Optimum storage temperature is 32 to 36° F at 95% relative humidity.

Quality Indices: Major quality indices that are considered for parsley crop are leaf colour, appearance, and crispiness. A description of the same is given below:

Quality Index	Description
Leaf colour	Deep green and bright
Appearance	Fresh and crispy; with long stalks; and free from yellowish, wilted, over mature, diseased and damaged leaves

Packing for Shipping: Hydrocooled (washed and cleaned) parsley leaves are bunched together into small bundles and placed in wooden boxes or corrugated carton boxes before shipping.

Processing: Freshly harvested parsley leaves are transported to the pack houses where it is washed and cleaned with water before drying it in a dark shaded place. Rapid drying is recommended in order to preserve the flavour and aroma of the produce. In rapid drying (quick drying) process, the leaves are subjected to oven drying at 95 Degree C (200 Degree F). In this method, the bright green color of the produce is retained. A major criterion is that ventilation should be provided during the entire drying process. Dried or dehydrated parsley leaves are called parsley leaf flakes and it is normally stored in dark storage chambers in air tight glass containers.

Hamburg Parsley (Turnip-Rooted Parsley)

In addition to curly-leaf parsley and Italian parsley, there is an edible root-forming variety of parsley which is called Turnip-rooted Parsley. Other common names of this parsley are Dutch parsley, Hamburg parsley, and Heimischer. Its scientific name is *Petroselinum crispum var. tuberosum*. Only white-colored root tubers of this parsley are edible and these tubers are cooked just like any other root vegetable such as carrot, turnip, parsnip, potato etc. These parsley roots have celery-like flavour and look like slender white carrots.

Growing practices for turnip-rooted parsley are similar to that of any other root vegetable. It is a short season crop and may take up to 3 to 4 months from seed sowing to harvesting. Tubers upon maturity are left in the ground until marketing, or may be dug and stored in underground cellars until marketing.

Peppermint

Peppermint is grown as an annual, aromatic herb for commercial purposes. It is mainly grown for the extraction of menthol, the active ingredient that imparts the characteristic aroma of peppermint oil. Peppermint, a hybrid mint variety, is native to the temperate regions of the European continent. It is a cross between water mint and spearmint. Scientific name of peppermint is *Mentha piperita*.

Peppermint is a long-stalked mint variety with opposite lanceolate leaves. It reaches up to 50 to 90 cm in height upon maturity. Smooth, square stems bear globular flower clusters in terminal spikes. Leaves are lanceolate, stalk-less and light to dark green in colour with reddish veins. Flowers are violet or purple in colour. Blooming time is from mid to late summer.

Growing Peppermint: A detailed account of various growing practices for peppermint plants is given below:

Requirements	Description
Climate	A cool, temperate and sub-temperate climate with a day temperature of 20 to 25°C is the most ideal temperature requirement. As far as rainfall is considered, a light rainfall favours vigorous growth of mint plants. Mint grows up to an altitude of 1,000 meters in temperate and subtropical regions. Generally speaking, wet climate and moist soils are the most ideal environmental conditions for growing almost all mint species.
Soil	Well-drained, rich sandy-loam to clay-loam soils with soil pH of 6–7 are the most ideal soils for growing peppermint. Clayey soils, high pH and frost are unsuitable.
Light	Peppermint is a long day plant, hence prefers sunny locations; partial shade is preferred at times, but full shade is undesirable.
Propagation	Being a hybrid, peppermint produces no seeds. Propagation is possible only by planting divided rhizomes (stolons or runners). Generally 8 to 10 cm long stolons with 2–4 growing points are used for propagation; suckers and herbaceous stem cuttings may also be used as propagation materials.
Planting Density	Approximately 250 kg/ha of fresh, healthy and succulent planting materials such as stolons or rhizomes or runners or cuttings.
Pre-treatment	Planting Materials should be treated with a recommended fungicide before planting them in the main field.
Spacing	Ideal spacing is 40 cm between two rows and 10 cm between two plants.
Planting Time	Mint can be planted any time of the year. However early spring is the best. Planting is done when soil temperature is low approx.@20°C
Planting Depth	0.5 cm – 1 cm deep in furrows
Field Preparation	The land is repeatedly ploughed twice or thrice; levelled and made free of all weeds before planting
Establishing Planting Materials	Cuttings or planting materials get established within 10 days

Manure and Fertilizer Application	In commercial cultivation practices, FYM (farm yard manure) or compost or vermicompost @ 15 - 30 tonnes/hectare along with NPK @ 40, 60 and 40 kg /hectare is mixed with the top soil at the time of land preparation; In Zinc-deficient soils, 20 kg of ZnSO should be mixed in the soil in order to avoid Zn deficiency. Nitrogen@80 Kg/hectare is also given as top dressing in 2 doses; first dose is 40 days after plants get established and second dose is after the appearance of the first flush.
Fertilizer Application	By broadcasting in rows
Irrigation	Mint is a shallow-feeder with shallow root system and hence deep water table with good drainage is recommended. In other words, mint is a moisture-loving plant and needs lots of moisture for its healthy growth. The crop needs 6–9 irrigations during dry season and 2–3 irrigations after rains.
Weed Control	Three weeding are required in total; each before every harvest; manual weed control is recommended
Disease-Pest Management	There is no severe insect-pest damage or disease incidence observed in mint plantation. Sometimes, mints are susceptible to whitefly and aphids. They can be controlled by spraying any of the recommended bioinsecticides such as neem emulsion or tobacco solution or soap solution.
Harvesting	Mint leaves may be harvested at any time for culinary purposes. For oil extraction, crop maturity needs to be determined by distillation of leaf sample in Clevenger's apparatus. If the average oil content in leaves is found to be around 0.5%, the crop is ready for harvesting. Generally speaking, mint crop for oil extraction is harvested at flowering stage. Harvesting during dry sunny days is recommended. 2-3 harvests in a year may be done. First harvest stage is reached in 105–110 days of planting. Subsequent harvests may be done at an interval of 80–90 days. During harvesting, fresh herbage is cut 10 cm above the ground using a sharp knife. Harvesting on cloudy or rainy day should be

	avoided as it decreases menthol content present in leaves.
Days to Maturity	240 days to complete a full growth cycle
Yield	Yield form commercial plantations: 15-20 tonnes fresh herbage/acre/annum in 2 harvests
Storage	*Storage of Fresh Leaves:* Fresh mint leaves are normally used immediately after harvest for culinary purposes such as garnishing. Alternatively, it may be stored up to 2-3 days in plastic bags in a refrigerator. Fresh mint leaves may also be stored for a few days by freezing in ice cube trays. *Storage of Dried Leaves:* leaves may be shade dried or air dried. Dried mint leaves can be stored in an airtight container placed in cool, dark, dry storages. Generally, essential oil content and aroma of mint leaves decrease during storage.
Drying and Distillation	*Air Drying under Shade:* Freshly harvested leaves and flowering tops are spread on a screen and dried in the shade. When the leaves and stems are brittle, excess stems are removed and dried leaves are cleaned before storing them in airtight containers. *Open Field Sun Drying:* Harvested fresh herbage is left in the field in the open sun for 4–6 hours for wilting until the fresh herbage loses 50% of its moisture. Dried herbage (dried leaves and flowering tops) is then transported to the distillation site where it is chopped into small pieces and distilled in a steam distillation unit
Yield of Oil	Approximately 100 litres of oil/acre
Oil Characteristics	The oil is golden-yellow in colour, sweet in taste with a pleasant, superior odour and is a mobile liquid. It contains 70–80 % menthol and menthone, the active principle behind cooling and gastro-stimulant properties of peppermint oil.

Uses of Pepper Mint Leaves and Oil

Both peppermint leaves and peppermint oil have several culinary, medicinal and commercial uses. Both fresh and dried peppermint leaves are used in culinary preparations. Peppermint oil is very high in menthol content. Menthol has high demand in food, medicinal and confectionary preparations. A description of various uses of peppermint is given below:

Culinary and Food Uses

1. Both fresh and dried leaves of peppermint is used as a garnishing agent in several food preparations such as fried rice, sauces, jellies, syrups, candies, soups and meat preparations
2. Fresh peppermint leaves are used for preparing special mint chutneys which aid in digestion after heavy spicy meals
3. Fresh peppermint leaves are used as a mouth freshener as they have a sweet aroma with a cool aftertaste
4. Fresh peppermint leaves are used in certain alcoholic drinks
5. Menthol, major ingredient in peppermint oil is extensively used as flavouring agent in certain beverages, mouth fresheners, toothpastes, ice creams and candies
6. Fresh peppermint leaves are used for preparing mint tea which is medicinal; relieves stomach ache
7. Peppermint oil is the most popular flavour of mint-flavoured confectionery items

Nutritive Value of Fresh Peppermint Leaves: Fresh peppermint leaves are rich sources of vitamins and minerals, particularly calcium, potassium, Vitamin A and Vitamin C. A detailed description of nutritive value of 100 gram edible portion of fresh peppermint leaves is given below:

Nutrient	Unit	Value per100g
Water	g	78.65
Energy	kcal	70
Protein	g	3.75
Total lipid (fat)	g	0.94
Carbohydrate, by difference	g	14.89
Fiber, total dietary	g	8
Calcium, Ca	mg	243
Iron, Fe	mg	5.08
Magnesium, Mg	mg	80
Phosphorus, P	mg	73
Potassium, K	mg	569
Sodium, Na	mg	31
Zinc, Zn	mg	1.11
Vitamin C, total ascorbic acid	mg	31.8
Thiamin	mg	0.082
Riboflavin	mg	0.266
Niacin	mg	1.706
Vitamin B-6	mg	0.129
Folate, DFE	µg	114
Vitamin B-12	µg	0
Vitamin A, RAE	µg	212
Vitamin A, IU	IU	4248
Fatty acids, total saturated	g	0.246
Fatty acids, total monounsaturated	g	0.033
Fatty acids, total polyunsaturated	g	0.508

Medicinal Uses

1. Peppermint oil is used as nervines, a medicinal preparation which works on the nervous system as a tranquilizer
2. Peppermint oil is analgesic, which alleviates pain
3. Peppermint oil is antispasmodic, which stops muscle spasms
4. Peppermint oil is carminative, which soothes the digestive system
5. Peppermint oil is laxative which clears congestion in the bowels

6. Peppermint oil is rubefacient, which improves skin blood circulation
7. Peppermint oil is anti-depressant which heals fatigue, stress, anxiety and depression
8. Fresh peppermint leaves are used to prepare mint tea which is used to treat stomach ache; mint tea is also a diuretic (promotes urine flow)
9. Menthol, major active ingredient in peppermint oil is used for the preparation of many medicinal drugs

Commercial Uses

1. Peppermint oil is used in aromatherapy
2. Menthol is an ingredient of many cosmetic products such as shampoos, soaps, skin care products, and perfumes
3. Products containing menthol are used as room fresheners
4. Peppermint oil has a high concentration of pulegone and menthone which can be used as natural pesticides
5. Peppermint flowers are rich in nectar and hence can be used for honey production

Spearmint

Spearmint is an erect-stemmed, vigorous-growing mint species with quadrangular purple-green, hairy stems. Scientific name of spearmint is *Mentha spicata*. This variety of mint is called spearmint because of its 'spear-shaped' or 'pointed' leaf tips. Flowering stems are narrow and long. Pink- or white-coloured flowers are borne on terminal, slender flowering spikes. The plant reaches up to a height of 60 to 100 cm upon full growth. Flowering time is from mid and late summer to early and mid fall.

Spearmint is mainly grown for its aromatic leaves which are used for culinary purposes as well as for oil extraction. Spearmint oil is rich in an aromatic compound called R-carvone which imparts the spearmint oil its characteristic aroma and flavour. Unlike peppermint, spearmint contains very less amounts of menthol and menthone.

Origin and Distribution: Spearmint is a native to the temperate regions of Europe and South West Asia. It is widely distributed throughout the temperate regions of North America.

Growing Spearmint: A detailed account of various growing practices for spearmint is given below:

Requirements	Description
Climate	Spearmint grows well in nearly all temperate climates
Soil	Spearmint is best suited to loamy soils with plenty of organic material. Soil pH requirements: 5.5 - 7.5
Light	The plant prefers partial shade; but can flourish both in full sun and shaded locations
Propagation	Spearmint does not set seed and its flowers are sterile. Propagation is done mainly by dividing rhizomes, from herbaceous stem cuttings and sometimes by planting suckers
Spacing	35-50 inches (90-120 cm)

Planting Time	Planting is done when temperature is around 20°C
Planting Depth	0.5 cm – 1 cm deep in furrows
Field Preparation	The land is repeatedly ploughed twice or thrice; levelled and made free of all weeds before planting
Fertilizer and Manure Application	In commercial cultivation practices, FYM (farm yard manure) or compost or vermicompost @ 15 - 30 tonnes/hectare along with NPK @ 40, 60 and 40 kg /hectare is mixed with the top soil at the time of land preparation; In Zinc-deficient soils, 20 kg of ZnSO should be mixed in the soil in order to avoid Zn deficiency. Nitrogen@80 Kg/hectare is also given as top dressing in 2 doses; first dose is 40 days after plants get established and second dose is after the appearance of the first flush. *Method of Application*: Organic manures are incorporated into the top soil at the time of land preparation. Nitrogenous fertilizers such as urea are broadcasted between the plant rows once plants are established.
Irrigation	Regular watering is needed until plants get established and thereafter watering is done as and when required. Overwatering and waterlogging should be avoided.
Weed Control	Three weeding in total, each before every harvest; manual weed control is recommended
Disease-Pest Management	There is no severe insect-pest damage or disease incidence observed in spearmint plantation
Harvesting	Spearmint leaves lose their aroma after the plant begins to flower and therefore harvesting is recommended just before flowering. Fresh leaves are harvested just before flowering stage on dry sunny days. 2-3 harvests in a year may be done. During harvesting, fresh herbage is cut 10 cm above the ground by using a sharp knife.
Yield	Yield of fresh herbage from commercial plantations ranges from 25 to 30 tons/ha and yield of oil is approximately 150 litres/ha

Drying and Distillation	According to the convenience of the grower, different containers such as plastic bags and cloth bags and different lighting conditions such as open sunlight and shaded dark chambers are used for drying spearmint leaves. *Open Field Sun Drying*: Harvested fresh herbage is left in the field in the open sun for 4–6 hours for wilting until the fresh herbage loses 50% of its moisture. *Air Drying under Shade*: Freshly harvested leaves are spread on a screen and dried in the shade. When the leaves and stems are brittle, excess stems are removed and dried leaves are cleaned before storing them in airtight containers. *Steam Distillation*: Dried herbage (dried leaves and flowering tops) is transported to the distillation site where it is chopped into small pieces and distilled in a steam distillation unit to extract spearmint oil.
Oil Characteristics	Spearmint oil is highly aromatic and is rich in R-carvone (up to 65%). Carvone is the active compound that gives the oil its distinctive medicinal and aromatic properties. Spearmint oil is rich in linalool also which provides its caraway-like odour.

Uses of Spearmint Leaves and Oil: Both spearmint leaves and spearmint oil have several culinary, medicinal and commercial uses. Both fresh and dried leaves are used in various culinary preparations. Spearmint oil is very high in carvone content and therefore has great demand in medical and cosmetic industries. A description of various uses of spearmint is given below:

Culinary and Food Uses: Fresh spearmint leaves are pungent with a pleasant aroma and flavour and hence popular among homemakers as a garnishing agent in various food preparations such as fried rice, spicy meat preparations, chutney and sauce preparations. Fresh spearmint leaves are an essential ingredient of famous Touareg Mint Tea which has great health benefits because of its high antioxidant activity. Spearmint tea is also a natural home remedy for stomach aches and mild digestive problems.

Fresh spearmint leaves are also a rich source of minerals such as iron (11.87mg/100g), calcium (199mg/100g), magnesium (63mg/100g), phosphorous (60mg/100g), sodium (30mg/100g) and potassium (458mg/100g) and vitamins such as Vitamin A (4054 IU/100g) and Vitamin C (13.3mg/100g). It is also rich in dietary fiber (6.8g/100g). A description of nutritive value of fresh spearmint is given below:

Nutrient	Unit	Value per 100g
Water	g	85.55
Energy	kcal	44
Protein	g	3.29
Total lipid (fat)	g	0.73
Carbohydrate, by difference	g	8.41
Fiber, total dietary	g	6.8
Calcium, Ca	mg	199
Iron, Fe	mg	11.87
Magnesium, Mg	mg	63
Phosphorus, P	mg	60
Potassium, K	mg	458
Sodium, Na	mg	30
Zinc, Zn	mg	1.09
Vitamin C, total ascorbic acid	mg	13.3
Thiamin	mg	0.078
Riboflavin	mg	0.175
Niacin	mg	0.948
Vitamin B-6	mg	0.158
Folate, DFE	µg	105
Vitamin B-12	µg	0
Vitamin A, IU	IU	4054
Fatty acids, total saturated	g	0.191
Fatty acids, total monounsaturated	g	0.025
Fatty acids, total polyunsaturated	g	0.394

Dried spearmint is richer and concentrated source of minerals, vitamins and dietary fiber. A description of nutritive value of dried spearmint is given below:

Nutrient	Unit	Value per 100g
Water	g	11.3
Energy	kcal	285
Protein	g	19.93
Total lipid (fat)	g	6.03
Carbohydrate, by difference	g	52.04
Fiber, total dietary	g	29.8
Calcium, Ca	mg	1488
Iron, Fe	mg	87.47
Magnesium, Mg	mg	602
Phosphorus, P	mg	276
Potassium, K	mg	1924
Sodium, Na	mg	344
Zinc, Zn	mg	2.41
Vitamin C, total ascorbic acid	mg	0
Thiamin	mg	0.288
Riboflavin	mg	1.421
Niacin	mg	6.561
Vitamin B-6	mg	2.579
Folate, DFE	Âµg	530
Vitamin B-12	Âµg	0
Vitamin A, IU	IU	10579
Fatty acids, total saturated	g	1.577
Fatty acids, total monounsaturated	g	0.21
Fatty acids, total polyunsaturated	g	3.257

At room temperature, fresh spearmint leaves lasts for only 2-3 days. It can be preserved for a long time by freezing. Salt, sugar or sugar syrup, alcohol, and oil may also be used for preserving fresh spearmint for a considerable period of time.

Medicinal Uses

1. Spearmint oil is used as carminative oil, which soothes the digestive system

2. Fresh spearmint leaves are used to prepare mint tea which is used to treat stomach ache

Commercial Uses

1. Spearmint oil is used as a flavouring agent for toothpastes and mouth cleaners
2. Spearmint oil is used as a base flavouring agent in many commercial scale food, beverage, and confectionery products
3. Spearmint oil is used in skincare products, perfumes, shampoos and soaps
4. Spearmint oil is used in aromatherapy

Japanese Mint

Scientific name of Japanese mint is *Mentha arvensis var. piperascens*. It is a vigorously growing, branched, hardy perennial plant which reaches up to a height of 1 meter upon maturity. Violet tinged quadrangular stems bear broadly ovate leaves; lilac-coloured flowers are borne in axillary and terminal flowering spikes.

Growing Japanese Mint: Growing practices for Japanese mint is similar to that of peppermint.

In fact, Japanese mint is mainly grown as a substitute of peppermint. Just like peppermint, Japanese mint is also grown for its aromatic leaves which are rich in essential oil, menthol. Just like peppermint oil, Japanese mint oil also contain up to 75–80% menthol and menthone content. Even though Japanese mint oil can be used as a substitute for peppermint oil, the latter has a superior odour and fetches higher price in the market than Japanese mint oil.

Bergamot Orange Mint

Scientific name of bergamot mint is *Mentha citrata*. Leaves of bergamot mint have a characteristic lemon-like or orange-like odour when crushed and therefore it is also known as orange mint or lemon mint. It is used for flavouring various drinks and cocktails.

Plant Description: It is a perennial aromatic herb that is robust in growth and grows up to 90 cm upon full growth. Leaves are broad and ovate but without a distinct inflorescence. The flowering vertices are borne in terminal axils of leaves. Flower colour is violet or lavender. Bloom time is late summer to early fall. It is a suitable mint variety for container gardening.

Growing Bergamot Orange Mint: A detailed account of various growing practices for bergamot orange mint is given below:

Requirements	Description
Climate	A cool, temperate and sub-temperate climate with a day temperature of 20–25°C is the most ideal requirement
Soil	Well-drained, rich sandy-loam to clay-loam soils are most ideal; Soil pH requirements: 6 to 8
Light	Prefers sunny locations and light shade
Propagation	Propagation is done by dividing the root

	ball (rhizomes); from herbaceous stem cuttings and by seeds. Seed propagation is possible. In seed propagation, seeds are allowed to dry on plants before collecting them.
Spacing	60-90 cm
Land Preparation and Fertilizer Application	As in case of other mint varieties
Irrigation, Weed Control, and Disease-Pest Management	As in case of other mint varieties
Harvesting, Drying and Distillation	As in case of other mint varieties
Yield of Fresh Herbage	15-20 tonnes fresh herbage/acre/annum in 2 harvests
Yield of Oil	Approximately 150 litres of oil/hectare
Oil Characteristics	The oil contains 45–50% linalool and up to 45% linalyl acetate

Uses of Bergamot Mint Leaves and Oil: Bergamot mint is extensively used for various medicinal and cosmetics applications. It is also used in perfumery industry. A detailed account of various medicinal uses of bergamot mint is given below:

Medicinal Uses of Bergamot Orange Mint

1. Bergamot mint oil is analgesic, which alleviates pain
2. Bergamot mint oil is antispasmodic, which stops muscle spasms
3. Bergamot mint oil is carminative, which soothes the digestive system
4. Bergamot mint oil is laxative which clears congestion in the bowels
5. Bergamot mint oil is rubefacient, which improves skin blood circulation

6. Mint tea is made from the fresh or dried bergamot mint leaves, which is a natural remedy for fevers, headaches, stomach aches, nausea and other digestive disorders.

Horse Mint

Horsemint is an herbaceous perennial plant mainly grown for its fresh foliage. The plants are with a peppermint-scented aroma. Stems are erect, sometimes creeping and stem reaches up to a height of 120 cm on full growth. Leaves are oblong-elliptical to lanceolate in shape with green colour on upper side and white colour below. Bloom time is mid to late summer and flower colour is lilac, purplish, or white. Flower clusters are produced on tall, branched, tapering spikes.

Origin and Distribution: Horsemint is believed to be a native of regions comprising of Europe, western and central Asia, and northern and southern Africa.

Growing Horsemint: Growing practices for horsemint is similar to that of other mint varieties.

Uses of Horsemint: Horsemint leaves are used in several culinary preparations. Horsemint has several medicinal applications also. Horse mint is used as an antiseptic, stimulant and is also useful in treating digestive complaints and fever.

Field Mint or Corn Mint

Corn mint is mainly grown for its aromatic foliage. It is an erect growing mint variety which reaches up to 60 cm in height upon full growth. It produces stalk less leaves with toothed leaf margins. Bloom time is mid-summer or late-summer to early fall or mid fall. Bloom color is pale pink.

Growing Practices: Growing practices for field mint is similar to that of other mint varieties.

Uses: Corn mint leaves are mainly used for culinary purposes as a garnishing and flavouring agent. It is also used in salads, sauces, soups, and meat preparations.

Apple Mint

Apple mint is *Mentha suaveolens*. It is popularly known as 'woolly mint' because the plant is covered with wool-like small hairs. Apple mint is an erect-growing herbaceous, perennial plant. It grows up to a height of 45-60 centimeter upon full growth. Leaves are light green in colour, oppositely arranged, hairy, wrinkled, oblong to nearly ovate in shape and with serrated leaf margins and rounded leaf tips. White or pinkish flowers are borne on terminal spikes. Bloom time is mid to late summer to early fall. This mint variety is suitable for growing in containers.

Origin and Distribution: Apple mint is native to southern and western Europe and the western Mediterranean region. It is naturalised in central and northern parts of Europe.

Growing Apple mint: Growing practices for apple mint is similar to that of other mint varieties.

Uses: Apple mint is mainly grown for its aromatic foliage which is used as a culinary herb. It is also used as an ornamental plant and as a ground cover. Fresh leaves are used to make apple mint jelly. It is used to make mint tea, as a garnish or flavouring agent and in soups and salads.

*Note: More detailed account of various types of mint herbs is available in my book titled "**Mint Herbs**"*

Asparagus

Asparagus is a cool-season vegetable that belongs to the family Liliaceae. Botanical name of asparagus is *Asparagus officinalis*. It is an herbaceous perennial plant. Tender shoots (spears) of asparagus plants are used as a leafy vegetable. Asparagus is believed to be originated in Europe, Africa and Asia. Young shoots of Asparagus are very rich in minerals like iron, phosphorus, potassium, copper, manganese, selenium, calcium, magnesium and zinc. It is also rich in vitamins and major vitamin present in asparagus are vitamin B6, vitamin A, vitamin C, vitamin E, vitamin K, thiamin, riboflavin, rutin, niacin, and folic acid. Asparagus shoots are low in calories, low in sodium, rich in protein and a rich source of dietary fiber. Asparagus shoots are also rich in an amino acid called asparagine. Dried asparagus roots are used as a medicine because of its diuretic properties. Productive life of a well-managed commercial asparagus plantation is about 10 to 15 years.

Cultivars of Asparagus: There are three types of cultivated asparagus: white asparagus; green asparagus and purple asparagus. Among these, green asparagus is largely consumed as a vegetable worldwide. Though white asparagus is less bitter than green asparagus, it is popular only in European countries like the Netherlands, France and Germany. Purple asparagus is commercially produced in Italy and it has high sugar and low fiber levels.

Soils for Asparagus Cultivation: Asparagus can be grown in many soil types, even in saline soils but deep loam or sandy soils with good surface water and air drainage are best. Good production is also possible in heavier soils. However, soil fertility may become a limiting factor for commercial cultivation of asparagus. Therefore a grower must ensure that soil is prepared well before planting the crowns in the main fields.

Soil pH: Asparagus plants grow well if the soil pH is within a range of 6.5 to 7.5. If soil pH is below 6.0 it should be raised to 6.5 by applying lime according to soil test report recommendations. Asparagus will thrive in soils having a salt content too high for many other crops, but it will not tolerate extreme acidity. Although asparagus will tolerate less than optimum soil conditions, yields are likely to be reduced and the life of the planting will be shortened in these soils.

Soil Drainage: In asparagus production, it is important that the plants develop an extensive storage root system. Therefore, good soil drainage is essential. Asparagus roots can develop to a depth of 10 ft in well-drained soils. It will do well in most soils if the water table does not come within three feet of the surface during the growing season.

Soil Preparation: Soil should be made fertile and free of troublesome weeds before the crowns are planted. Since it is more difficult to improve soil after the crowns are planted, soil improving practices must be started at least a year before planting. Asparagus thrives best in soils well supplied with organic matter. Applications of animal manure or turning under a green manure crop are desirable practices prior to planting asparagus. Green manure crops also improve soil structure and enhance soil fertility. Phosphorous, potassium, and lime amendments—based on a soil test—may also be incorporated prior to planting.

Propagation: Asparagus can be propagated both by seeds and by crowns. For small plantations and home gardens, it is advised to purchase one-year-old crowns from a reliable grower or nurseryman. For large commercial plantations, an asparagus producer can raise crowns by sowing seeds in nursery beds.

Raising Crowns from Seeds

Nursery bed preparation: In order to grow crowns for a commercial plantation, high quality seeds should be purchased from a reliable nursery and then planted in well prepared nursery beds that have never grown asparagus before. Sandy soils are recommended for nursery bed preparation so that crowns can be easily dug from the nursery beds. Based on soil testing report, NPK (Nitrogen, Phosphorous and Potassium) fertilizers should be incorporated into the soil prior to sowing the seeds.

Seed Sowing Time: Normally in tropics, seed are planted in early April

Spacing: Seeds are sown in rows which are kept 2 to 3 feet apart. A planting depth of one to two inches is needed.

Seed Rate: 400-500 grams of seeds is required to produce crowns for planting one acre of asparagus.

Germination: Asparagus is slow to germinate requiring two to three weeks for the seedlings to emerge. One-year old seedlings are used for transplanting.

Handling Crowns: While lifting the asparagus crowns from the nursery rows, care should be taken not to injure the crowns during digging. Crowns should be planted in the main field as soon as possible after digging.

Setting the Crowns: Before planting the crowns in the main field, the individual crowns that are grown together are separated and prepared well by removing unhealthy leaves and roots.

Planting Asparagus Crowns in the Main Field

Field Preparation: Main fields are prepared by ploughing and levelling and by incorporating organic manures into the top soil. Since asparagus is a perennial crop and the plants occupy the soil for several years it is essential that the soil should be enriched by incorporating vast amounts of organic manures in the form of farm yard manure or compost or green manure crops into the top soil. After soil is set, furrows which are four to six inches deep are prepared for planting the crowns.

Planting Crowns: Crowns are placed in the bottom of furrows carefully and then the crowns are covered with two to three inches of soil. A light irrigation is recommended soon after planting.

Spacing: Rows are spaced five to six feet apart, and plants are spaced 12 to 15 inches apart in the row.

Plant Population per Unit Area: Recommended plant population is 6,000 to 9,000 crowns per acre for the given spacing in and between rows.

Planting Time: In the tropics, asparagus crowns are normally planted at the onset of winter or during the winter months. In such cases, first shoots start appearing during spring season.

Fertilizer Application

Fertilizer Application during the First Year: Based on the soil test analysis, required amounts of NPK fertilizers should be supplied to the soil as basal dressing before planting the crowns in the main field.

Fertilizer Application of Established Plants: Recommended NPK fertilizer dose for acreage of established asparagus plantation in tropics is an annual dose of 50-60 Kg of nitrogen (N), 25 Kg of phosphorous (P), and 50 Kg of potassium (K) in two applications, once in spring and again after harvesting.

Irrigation: Asparagus is a drought-resistant crop. Since asparagus has an extensive and deep root system, frequent irrigation is not necessary. Irrigation should be done as and when necessary.

Insect Pests: Major insect pests that are found affecting an asparagus plantation are, asparagus beetle, spotted asparagus beetle, and asparagus aphid.

Control Measures

1. Field sanitation by burning all the trash and debris found at the growing site; these are the places where the beetles are likely to hibernate
2. Application of biopesticides such as Rotenone or rotenone-pyrethrum mixtures
3. By using natural predators of beetles such as chalcid wasps and lady beetle larvae
4. By using insecticidal soap solution and rotenone-pyrethrum sprays

Diseases: Major fungal diseases that are found affecting an asparagus plantation are asparagus rust, needle blight, crown rot, root rot and wilting of plants. Major symptom of rust disease is defoliation of leaves and in needle blight disease, leaves get purple-brown spots and are slowly defoliated. Needle blight is caused by *Cercospora asparagi*. Wilting of the plants is caused by the fungal species *Fusarium* and asparagus rust is caused by *Puccinia asparagi*.

Control Measures: Asparagus rust can be controlled by growing rust-resistant varieties. Other fungal diseases may be controlled by using organic fungicidal solutions.

Weeds and Weed Control: Asparagus plantation is found affected by both annual and perennial weeds. Weeds can effectively be controlled by adopting a combination of cultural, mechanical, biological and chemical weed control techniques. In cultural weed control techniques, a combination of tillage, mulching, soil solarization and cover cropping may be adopted at least a year before planting crowns in the field to ensure a weed-free site for plantation.

Harvesting: Asparagus should not be harvested during the first two years of planting. In the year of setting crowns, the plants should be allowed to get established in the field and during the second year plants should be allowed to grow vigorously and develop a strong storage root system. From third year of planting onwards, tender shoots, which are also called 'spears' may be harvested. Harvest is done when there are three to four flushes spear emergence. In an established plantation, harvest season may be extended to eight to nine weeks. Approximately, two-thirds of the crop is harvested during the first half of the harvest season itself. Asparagus spears are harvested when they are about 9-10 inches in length. Spears are either handpicked by snapping them just above the ground or cut by a special knife just below the ground.

Shelf Life: Shelf life of freshly harvested asparagus spears is up to one week to ten days at room temperatures.

Economic Life of an Asparagus Plantation: A well-established asparagus plantation starts yielding sizeable returns after about three years. Under good cultural management conditions, a plantation may remain economical for about 12 years.

Cost of Production: The major expense in asparagus production is the initial cost to establish the plantation. These are,

1. Cost of seeds or crowns
2. Cost of organic manures, fertilizers and plant chemicals
3. Labor-related costs – site preparation, fertilizer and plant chemical application, irrigation, harvesting etc
4. Cost of maintaining the plantation: Cost of maintaining the crop after establishment is less than costs incurred in producing most other vegetable crops.

Pre-Cooling: Freshly harvested asparagus spears are pre-cooled by rapid hydro cooling. Hydrocooling is accomplished by flooding, spraying, or immersing asparagus spears in chilled water. Following hydrocooling, asparagus should be kept refrigerated.

Quality Indices: Major quality indices of asparagus spears are given below in a tabular format.

Quality index parameter	Description
Appearance	Fresh
Color	Dark Green
Texture	Firm and with Tightly Closed, Compact Tips
Stalks	Glossy, Straight, and Tender

Grades: Major market grades of asparagus spears are given below in a tabular format.

Grade	Size of the Spears (length)
Standard	9-10 inches

Optimum Temperature: Recommended optimum temperature is 0°C-2°C (32°F-35.6°F). At 2°C, storage life is two to three weeks and at 0°C, storage life can be prolonged up to a month.

Optimum Relative Humidity: Recommended optimum humidity is 95-100%.

Physiological and Physical Disorders: A detailed account of various disorders in asparagus plants is given below:

Bending: Upward bending of tips away from gravity and "feathering" (expansion and opening) of tips is a common physiological order found in asparagus.

Spear toughening: It occurs at a temperature above 10°C (50°F).

Bruising: Bruising and tip-breakage are signs of rough handling which often result in spear toughening from wound ethylene.

Chilling Injury: It occurs when spears are kept at 0°C (32°F) for more than 10 days. Affected spears lose glossiness and look wilted in appearance. Severe chilling injury results in darkening near tips in spots.

Freezing injury: It occurs when spears are stored at temperatures of -0.6°C (30.9°F) or lower. Symptoms include water-soaked appearance leading to extreme softening.

Pathological Disorders: Most common disease is Bacterial soft rot, caused by *Erwinia carotovora subsp. carotovora*. In this disease, spear tips start decaying.

Standard Packaging: Asparagus spears are packaged with water-soaked pads to maintain turgidity of the spears.

Packaging for Shipment: Pyramid-shaped wooden or waxed corrugated boxes are used for packaging spears which combined with center-loading during shipment promote good cooling-air circulation.

Nutritional Information for Fresh and Processed Asparagus

Food Uses: Asparagus spears are a major ingredient in many European and Thai cuisines. They are used in stew preparations, pickles, and for making delicious vegetable preparations. Nutritional information of raw asparagus spears as well as various asparagus food preparations is given below:

Nutrition in raw (fresh) asparagus spears

Nutrient	Unit	Value/ 100g	1.0"cup"134.0g	1.0"spear, small (5" long or less)"12.0g	1.0"spear, medium (5-1/4" to 7" long)"16.0g	1.0"spear, large (7-1/4" to 8-1/2")"20.0g	1.0"spear, extra large (8-3/4" to 10" long)"24.0g	1.0"spear tip (2" long or less)"3.5g
Water	g	93.22	124.91	11.19	14.92	18.64	22.37	3.26
Energy	kcal	20	27	2	3	4	5	1
Protein	g	2.2	2.95	0.26	0.35	0.44	0.53	0.08
Total lipid	g	0.12	0.16	0.01	0.02	0.02	0.03	0
Carbohydrate	g	3.88	5.2	0.47	0.62	0.78	0.93	0.14
Fiber	g	2.1	2.8	0.3	0.3	0.4	0.5	0.1
Sugars	g	1.88	2.52	0.23	0.3	0.38	0.45	0.07
Calcium	mg	24	32	3	4	5	6	1
Iron, Fe	mg	2.14	2.87	0.26	0.34	0.43	0.51	0.07
Magnesium	mg	14	19	2	2	3	3	0
Phosphorus, P	mg	52	70	6	8	10	12	2
Potassium, K	mg	202	271	24	32	40	48	7
Sodium, Na	mg	2	3	0	0	0	0	0
Zinc	mg	0.54	0.72	0.06	0.09	0.11	0.13	0.02
Vitamin C	mg	5.6	7.5	0.7	0.9	1.1	1.3	0.2
Thiamin	mg	0.143	0.192	0.017	0.023	0.029	0.034	0.005
Riboflavin	mg	0.141	0.189	0.017	0.023	0.028	0.034	0.005
Niacin	mg	0.978	1.311	0.117	0.156	0.196	0.235	0.034
Vitamin B-6	mg	0.091	0.122	0.011	0.015	0.018	0.022	0.003
Folate, DFE	Åµg	52	70	6	8	10	12	2
Vitamin B-12	Åµg	0	0	0	0	0	0	0
Vitamin A, IU	IU	756	1013	91	121	151	181	26
Vitamin E	mg	1.13	1.51	0.14	0.18	0.23	0.27	0.04
Vitamin K	Åµg	41.6	55.7	5	6.7	8.3	10	1.5

Nutrition in frozen, cooked, boiled, drained (without salt) asparagus

Nutrient	Unit	Value per100g	1.0"cup "180.0g	1.0"package (10 oz) yields"293.0g	4.0"spears"60.0g
Water	g	94.1	169.38	275.71	56.46
Energy	kcal	18	32	53	11
Protein	g	2.95	5.31	8.64	1.77
Total lipid (fat)	g	0.42	0.76	1.23	0.25
Carbohydrate	g	1.92	3.46	5.63	1.15
Fiber, total dietary	g	1.6	2.9	4.7	1
Sugars, total	g	0.32	0.58	0.94	0.19
Calcium, Ca	mg	18	32	53	11
Iron, Fe	mg	0.56	1.01	1.64	0.34
Magnesium, Mg	mg	10	18	29	6
Phosphorus, P	mg	49	88	144	29
Potassium, K	mg	172	310	504	103
Sodium, Na	mg	3	5	9	2
Zinc, Zn	mg	0.41	0.74	1.2	0.25
Vitamin C	mg	24.4	43.9	71.5	14.6
Thiamin	mg	0.065	0.117	0.19	0.039
Riboflavin	mg	0.103	0.185	0.302	0.062
Niacin	mg	1.038	1.868	3.041	0.623
Vitamin B-6	mg	0.02	0.036	0.059	0.012
Folate, DFE	Âµg	135	243	396	81
Vitamin B-12	Âµg	0	0	0	0
Vitamin A, IU	IU	806	1451	2362	484
Vitamin E	mg	1.2	2.16	3.52	0.72
Vitamin K	Âµg	80	144	234.4	48
Fatty acids, total saturated	g	0.096	0.173	0.281	0.058
Fatty acids, total monounsaturated	g	0.013	0.023	0.038	0.008
Fatty acids, total polyunsaturated	g	0.185	0.333	0.542	0.111

Nutrition in frozen asparagus (cooked, boiled, drained with salt)

Nutrient	Unit	Value per 100g	1.0"cup"180.0g	1.0"package (10 oz) yields"293g	4.0"spears"60.0g
Water	g	94.1	169.38	275.71	56.46
Energy	kcal	18	32	53	11
Protein	g	2.95	5.31	8.64	1.77
Total lipid (fat)	g	0.42	0.76	1.23	0.25
Carbohydrate	g	1.92	3.46	5.63	1.15
Fiber, total dietary	g	1.6	2.9	4.7	1
Sugars, total	g	0.32	0.58	0.94	0.19
Calcium, Ca	mg	18	32	53	11
Iron, Fe	mg	0.56	1.01	1.64	0.34
Magnesium	mg	10	18	29	6
Phosphorus, P	mg	49	88	144	29
Potassium, K	mg	172	310	504	103
Sodium, Na	mg	240	432	703	144
Zinc, Zn	mg	0.41	0.74	1.2	0.25
Vitamin C	mg	24.4	43.9	71.5	14.6
Thiamin	mg	0.065	0.117	0.19	0.039
Riboflavin	mg	0.103	0.185	0.302	0.062
Niacin	mg	1.038	1.868	3.041	0.623
Vitamin B-6	mg	0.02	0.036	0.059	0.012
Folate, DFE	µg	135	243	396	81
Vitamin B-12	µg	0	0	0	0
Vitamin A, IU	IU	806	1451	2362	484
Vitamin E	mg	1.2	2.16	3.52	0.72
Vitamin K	µg	80	144	234.4	48
Fatty acids, total saturated	g	0.096	0.173	0.281	0.058
Fatty acids, total monounsaturated	g	0.013	0.023	0.038	0.008
Fatty acids, total polyunsaturated	g	0.185	0.333	0.542	0.111

Nutrition in canned (drained solids) asparagus: A detailed account of nutrition in canned asparagus is given below:

Nutrient	Unit	Value per100g	1.0"cup"242.0g	1.0"spear (about 5" long)"18.0g	1.0"can (300 x 407)"248.0g
Water	g	93.98	227.43	16.92	233.07
Energy	kcal	19	46	3	47
Protein	g	2.14	5.18	0.39	5.31
Total lipid	g	0.65	1.57	0.12	1.61
Carbohydrate	g	2.46	5.95	0.44	6.1
Fiber	g	1.6	3.9	0.3	4
Sugars	g	1.06	2.57	0.19	2.63
Calcium	mg	16	39	3	40
Iron, Fe	mg	1.83	4.43	0.33	4.54
Magnesium	mg	10	24	2	25
Phosphorus, P	mg	43	104	8	107
Potassium, K	mg	172	416	31	427
Sodium, Na	mg	287	695	52	712
Zinc, Zn	mg	0.4	0.97	0.07	0.99
Vitamin C	mg	18.4	44.5	3.3	45.6
Thiamin	mg	0.061	0.148	0.011	0.151
Riboflavin	mg	0.1	0.242	0.018	0.248
Niacin	mg	0.954	2.309	0.172	2.366
Vitamin B-6	mg	0.11	0.266	0.02	0.273
Folate, DFE	µg	96	232	17	238
Vitamin B-12	µg	0	0	0	0
Vitamin A, IU	IU	822	1989	148	2039
Vitamin E	mg	1.22	2.95	0.22	3.03
Vitamin K	µg	41.3	99.9	7.4	102.4
Fatty acids, total saturated	g	0.147	0.356	0.026	0.365
Fatty acids,total monounsaturated	g	0.021	0.051	0.004	0.052
Fatty acids, total polyunsaturated	g	0.284	0.687	0.051	0.704

Nutrition in cooked (boiled and drained) asparagus: A detailed account of nutrition in COOKED asparagus is given below:

Nutrient	Unit	Value per100g	0.5"c up"90 .0g	4.0"spears (1/2" base)"60.0g
Water	g	92.63	83.37	55.58
Energy	kcal	22	20	13
Protein	g	2.4	2.16	1.44
Total lipid (fat)	g	0.22	0.2	0.13
Carbohydrate	g	4.11	3.7	2.47
Fiber, total dietary	g	2	1.8	1.2
Sugars, total	g	1.3	1.17	0.78
Calcium, Ca	mg	23	21	14
Iron, Fe	mg	0.91	0.82	0.55
Magnesium, Mg	mg	14	13	8
Phosphorus, P	mg	54	49	32
Potassium, K	mg	224	202	134
Sodium, Na	mg	14	13	8
Vitamin C	mg	7.7	6.9	4.6
Thiamin	mg	0.162	0.146	0.097
Riboflavin	mg	0.139	0.125	0.083
Niacin	mg	1.084	0.976	0.65
Vitamin B-6	mg	0.079	0.071	0.047
Folate, DFE	Âµg	149	134	89
Vitamin B-12	Âµg	0	0	0
Vitamin A, IU	IU	1006	905	604
Vitamin E	mg	1.5	1.35	0.9
Vitamin K	Âµg	50.6	45.5	30.4
Fatty acids, total saturated	g	0.048	0.043	0.029
Fatty acids, total polyunsaturated	g	0.105	0.094	0.063

Nutrition in frozen (unprepared) asparagus: A detailed account of nutrition in FROZEN asparagus is given below:

Nutrient	Unit	Value/ 100g	4.0"spears" 58.0g	1.0"package (10 oz)"284.0g
Water	g	91.82	53.26	260.77
Energy	kcal	24	14	68
Protein	g	3.23	1.87	9.17
Total lipid (fat)	g	0.23	0.13	0.65
Carbohydrate	g	4.1	2.38	11.64
Fiber, total dietary	g	1.9	1.1	5.4
Calcium, Ca	mg	25	14	71
Iron, Fe	mg	0.73	0.42	2.07
Magnesium, Mg	mg	14	8	40
Phosphorus, P	mg	64	37	182
Potassium, K	mg	253	147	719
Sodium, Na	mg	8	5	23
Zinc, Zn	mg	0.59	0.34	1.68
Vitamin C	mg	31.8	18.4	90.3
Thiamin	mg	0.121	0.07	0.344
Riboflavin	mg	0.131	0.076	0.372
Niacin	mg	1.202	0.697	3.414
Vitamin B-6	mg	0.111	0.064	0.315
Folate, DFE	Âµg	191	111	542
Vitamin B-12	Âµg	0	0	0
Vitamin A, IU	IU	948	550	2692
Fatty acids, total saturated	g	0.052	0.03	0.148
Fatty acids, total monounsaturated	g	0.007	0.004	0.02
Fatty acids, total polyunsaturated	g	0.101	0.059	0.287

Chives

Scientific name of Chives is *Allium schoenoprasum*. It is a bulbous herbaceous, perennial leafy vegetable belonging to the family of onions, i.e. Amaryllidaceae.

Taxonomy

Kingdom Plantae
Class Angiosperms/Monocots
Order Asparagales
Family Amaryllidaceae
Genus Allium
Species A. schoenoprasum

Origin and Distribution: Chives are believed to be a native to European nations, particularly Britain. It is also found naturally growing throughout Asian continent and North America

Nutritional Information: Chives are rich in vitamins A and C and also contain trace amounts of sulfur, and are rich in calcium and iron.

Botanical Description: A detailed description of this plant is given below:

Plant: A chive plant grows up to an average height of 1 to 1.5 feet and an average spread of 1 to 1.5 feet; a mature plant produces underground bulbs which are slender conical in shape. An average bulb is 3 centimetres in length and one centimetre in breadth. Plant stems are hollow and tubular with each stem measuring up to 3 millimetres in diameter.

Flowers: The chive plant produces attractive, showy purple or pink flowers in thick umbel inflorescences with each inflorescence carrying up to 30 flowers in a cluster. Each flower is star-shaped with six petals with approximately one centimetre in width. Flowering time is from April to June.

Fruits: Fruit is a capsule which contains numerous tiny seeds in three valves. Fruit matures in summer season. Seeds are small and black in color.

Uses: A detailed account of various food uses of chives is given below:

Culinary Uses: Chives are used as a flavouring herb. Fragrant leaves of the chive plants are the part primarily used in food preparations. Due to their strong flavours and aromas, chives are chopped into fine pieces and used in small quantities. Chives are a common ingredient in Chinese, Japanese and Korean food preparations. Chives are used as an ingredient for fish preparations and also used with potatoes. Finely chopped chives are used as a garnish in soups.

Uses as an Insect-Repellent: Chive plants are known for their insect-repelling properties due to the presence of sulphur compounds in their cells, and therefore popularly cultivated in home gardens to control insects and pests

Uses as a Fungicide: Juice extracted from the leaves can be used as a fungicide to fight fungal infections such as mildew and scab

Medicinal Uses: Chives may be used as a mild stimulant, diuretic, and an antiseptic and also have a beneficial effect on the circulatory system due to the presence of allyl sulfides and alkyl sulfoxides.

Ornamental Uses: Chives may be used as an edging plant in gardens because they are easy to grow and have attractive flowers. Purple- violet flowers of chives are used in ornamental dry bouquets.

Production Practices: Chives are easy to grow drought-resistant perennial plants which can be grown as annuals as well.

Soil Requirements: Chive plants prefer well drained soil, which is highly fertile and is rich in organic matter, with a pH of 6-7. They should be grown under full sunlight for their healthy growth.

Propagation: Chives can be propagated through seeds and also by division of clumps.

Planting Time: Chive seedlings are planted during summer, or early the following spring

Raising Seedlings from Seeds: Seeds are sown in nursery beds and then beds are covered with appropriate biodegradable mulch. Seed beds are always kept moist until seeds start germinating. Under normal management practices, chive seeds germinate within a fortnight if temperature is at 15 °C to 20 °C.

Transplanting Seedlings in the Main Fields: Four week-old seedlings are ready for transplanting in the main fields.

Irrigation Requirements: Light irrigation is done at the time of planting. Thereafter watering is done as and when required. Soil should always be moist but overwatering should be avoided.

Disease-Pest Management: No serious insect or disease problems are found in chives. However root rot may occur in overwatered or wet, poorly-drained soils.

Harvesting of Chive leaves: Since leaves of chive plants are used as a leafy vegetable, harvesting is done by cutting down the required number of stalks from their base. Since plants continuously regrow very soon, continuous harvesting is possible. i.e. 3-4 harvests in a growing season.

Economic Life of a Chives Plantation: Once planted, the chives can be harvested at least for three to four years before replanting the field with new plants.

Post Harvest Practices: A detailed account of post harvest practices for chives is given below:

Quality Indices

Appearance	✓ Fresh and green leaves ✓ No yellowing ✓ No decay ✓ No insect damage ✓ No mechanical damage ✓ Uniform sized ✓ Strong flavour and aroma
Grades and Sizes	There are no market grades or sizes for chives

Packaging: Uniform sized bunches are tied together with rubber bands, and then packaged in plastic bags or clamshell containers before placing them in standard-sized corrugated cartons. Perforated polyethylene liners should be used in the cartons.

Pre-cooling: Precooling is done at or above 0 °C (32 °F) immediately after harvest

Optimum Storage Conditions: Chives are stored for 2 to 3 weeks at 0 °C (32 °F) with 95 to 100% relative humidity

Controlled Atmosphere (CA) Storage: Due to a short postharvest life, CA storage is not recommended for chives

Retail Outlet Display: Use of water sprinklers is acceptable to maintain the freshness of freshly arrived chives

Chilling Sensitivity: Chives are not chilling sensitive

Ethylene Production and Sensitivity: Ethylene production is low, but ethylene sensitivity is high. Storage at 0 °C reduces adverse effects of ethylene on visual quality

Respiration Rates at 0 °C: 22 mg CO_2 kg-1 h-1

Physiological Disorders: Yellowing and leaf abscission due to ethylene exposure are major physiological disorders found in chives.

Postharvest Pathology: Molds and bacterial decay are some major issues during the storage of chives. Maintenance of low temperatures during storage and distribution may slow down the rate of decay

Processing: Drying of chive leaves is not recommended as it will lose natural flavour of the leaves. If chives are to be stored for long term, freezing is recommended. Chive leaves are chopped into fine pieces and stored in ice cube trays with water before freezing them.

Major Markets for Chives: Chives are available in the market throughout the year. USA, France, Sweden, Poland and other European countries have a good market for chives.

Onion

Onion is one of the most popular vegetables in the world. It is an essential ingredient in almost all Asian food preparations. Onion is believed to be originated in the region comprising of India, Afghanistan and the Soviet Republics.

Types of Onions: Based on colour of the skin, there are three types – brown to red onions, yellow onions and white onions.

Climate for Growing Onions: Onion is a cool season crop which is subtropical to tropical in its growth habit. High temperature and long photoperiod (i.e. long days) are essential for successful growing of onions. Seed stalk development is completely dependent on the availability of high temperature and long photoperiod.

Soil for Growing Onions: Onions may be grown in all types of soils. However, the best soil is that is rich in humus and organic matter. High soil acidity needs to be avoided as onion plants are highly sensitive to soil acidity. Optimum pH range is between 5.8 and 6.5.

Propagation: Seed propagation is practiced in onions. Seedlings are raised in a well-prepared nursery bed. Transplanting is done when seedlings are about one or two months old.

Site Preparation for Growing Onions: Main site is prepared by ploughing, levelling and by incorporating organic manures and fertilizers into the top soil. Generally, 10 to 20 metric tons of farm yard manure is incorporated into the top soil to enhance soil fertility. Ridges are prepared at 3 meter distance and seedlings are transplanted on these ridges at 10-15 centimetres apart.

Nutrient Requirement: Researches reveal that a yield of 300 kilos of onions removes approximately 150 grams of nitrogen, 420 grams of phosphate and 1300 grams of potash.

Nutrient Application: Half dose of nitrogen and full doses of phosphate and potash fertilizers are applied along with farm yard manure. Rest half of the nitrogen is given one month after transplanting.

Weeding: Manual weeding is recommended

Irrigation: Irrigation frequency depends on climatic conditions and soil moisture level. It is important that soil moisture needs to be kept at optimum level throughout all growth stages of the onion plant. Irrigation needs to be stopped when onion tops start maturing and showing symptoms of wilting.

Harvesting Stage or Maturity Index: Harvesting is done when approximately 10 to 20 percent of tops have fallen over.

According to UC Davis, "Field-dry" maturity is indicated when bulb neck is completely dry to the touch and not slippery. This stage is typically reached at 5-8% weight loss following harvest.

Quality Indices

1. Mature neck and scales
2. Firmness

3. Diameter (Bulb size)
4. Absence of decay, insect damage, sunscald, greening, sprouting, freezing injury, bruising, and other defects
5. Degree of pungency

Yield: One acre (i.e. 4000 square meter area) yields approximately 15 tons of onion bulbs.

Curing: Harvested onion bulbs are exposed to sunlight for 3-4 weeks in order to be cured before storing them in containers. According to UC Davis, *field curing is done when temperatures are at least 24°C (75°F) or exposure for 12 hrs to 30 to 45°C (86 to 113°F) for forced air-curing.*

Storage Life: Storage life for mild onions and pungent onions is given below:

Mild onions	Typically 0.5 to 1 month at 0°C (32°F)
Pungent Onions	Typically up to 6 to 9 months at 0°C (32°F) depending on the cultivar

Optimum Relative Humidity: According to UC Davis, optimum RH is, *75 to 80% for best scale color development and for storage, 65 to 70% with adequate air circulation (1m3/min/m3 of onion).*

Responses to Ethylene: According to UC Davis, *ethylene may encourage sprouting and growth of decay-causing fungi.*

Physiological and Physical Disorders: Various physiological and physical disorders in onions are as below:

Disorder	Symptom
Freezing Injury	Soft water-soaked scales
Translucent Scales	Resembles freezing injury Control: cold storage following curing
Greening	Green-coloration of outer scales
Ammonia Injury	Brown-black blotches

Pathological Disorders: Various pathological disorders of onions are as below:

Disorder	Symptom
Botrytis Neck Rot	Watery-decay at neck area and light gray to gray fungal growth at neck infection and on outer scales Control: Proper drying and curing of onion
Black Mold	It is a fungal infection Causal organism is *Aspergillus niger* Symptom: black discoloration and shrivelling at neck and on outer scales Control: Low temperature storage following field or handling infestation
Blue Mold	It is a fungal infection Causal organism is *Pencillium spp.* Symptom: Watery soft rot of neck and outer scales followed by the appearance of green-blue mold fungal spores Control: Minimize mechanical injury, sunscald, and freezing injury
Bacterial Rots/Soft Rot	It is a bacterial infection Causal organism is *Erwinia carotovora subsp. Carotovora* Symptom: Water-soaked, foul-smelling, viscous liquidy rot Control: Harvesting at full maturity; proper drying and curing; minimizing mechanical injury and maintaining proper storage conditions
Slippery Skin	Symptom: slippery skin at neck area and on inner scales which have a watery-cooked appearance
Sour Skin	Symptom: slimy, yellow-brown decay on inner scales with a sour odor

Garlic

Garlic is one of the most commercially important vegetables because of its various medicinal properties. Garlic is believed to be originated in Central Asia.

Growing Garlic: A detailed account of various growing practices for garlic plants is as given below:

Climate: Garlic is a cool season crop which is subtropical to tropical in its growth habit. Garlic grows best where onions grow best.

Soil: Soil requirements for garlic growing are same as that of onions.

Propagation: Propagation is via cloves which are detached individually from the garlic bulb.

Clove Requirement: Approximately 250 individual cloves are required to plant an area of one acre (i.e. 4000 square meter).

Site Preparation: Main site is prepared by ploughing, levelling and by incorporating organic manures and fertilizers into the top soil. Generally, 10 to 20 metric tons of farm yard manure is incorporated into the top soil to enhance soil fertility.

Spacing: Rows are prepared at 15 centimeter distance and cloves are planted at 7.5 centimetres within the rows.

Nutrient Requirement: Nutrient requirement is similar to that of onions. Garlic may need more nitrogen.

Weeding: Manual weeding is recommended

Irrigation: Irrigation frequency depends on climatic conditions and soil moisture level. It is important that soil moisture needs to be kept at optimum level throughout all growth stages of the onion plant. Irrigation needs to be stopped when onion tops start maturing and showing symptoms of wilting.

Harvesting Stage and Maturity Index: Garlic is harvested when the bulbs are well mature. Harvest occurs after the tops have fallen and are very dry. (*Source: UC Davis*)

Yield: One acre (i.e. 4000 square meter area) yields approximately 2000 to 4000 kg of garlic cloves.

Quality Indices

Quality Index	Description
Appearance	Clean and White
Curing	Dried Neck and Outer Skins are Properly Cured
Cloves	Firm to the Touch; a High Dry Weight; Soluble Solids Content should be >35%

Grading: Major grading parameters are external appearance and diameter of each individual clove. External appearance should be clean, white and free from any external defects. Diameter of individual clove needs to be about 4 centimeters i.e. 1.5 inches for fresh market. Major two grades are *U.S. No. 1* and *Unclassified*.

Curing: Harvested bulbs are exposed to sunlight for 3-4 days in order to be cured before storing them in containers. According to UC Davis, *curing of garlic is the process by which the outer leaf sheaths and neck tissues of the bulb are dried. Warm temperatures, low relative humidity, and good airflow are conditions needed for efficient curing.* Garlic curing is normally done in the field. Curing is essential to obtain maximize storage life and have minimal decay.

Optimum Temperature and Relative Humidity:
Recommended optimum temperature (for best long term storage for over 9 months) is **-1°C to 0°C (30°F-32°F)** with low relative humidity (60-70%) and good ventilation. However, according to UC Davis, *garlic can be kept in good condition for 1-2 months at ambient temperatures i.e. 20°C-30°C [68°F-86°F] under low relative humidity (<75%).*

Physiological and Physical Disorders: A detailed account of various physiological disorders in garlic is as given below:

Disorder	Symptom
Freeze injury	Garlic freezes at temperatures below -1°C (30°F)
Waxy breakdown	Cause: high temperature near harvest
	Symptom: Small, light yellow areas in the clove flesh that darken to yellow or amber with time
	Clove is translucent, sticky and waxy
	Outer dry skins are not usually affected

| | Found in stored and shipped garlic |
| | Low oxygen levels and inadequate ventilation during handling and storage |

Pathological Disorders: A detailed account of various pathological disorders in garlic is as given below:

Disorder	Symptom
Penicillium rot	Causal organism: Pencillium corymbiferum and other spp Symptom: Affected garlic bulbs are light in weight and the individual cloves are soft and spongy and powdery dry. In an advanced stage of decay, the cloves break down in a green or gray powdery mass. Low humidity in storage retards rot development
Fusarium basal rot	Causal organism: Fusarium oxysporum cepae Symptom: It infects the stem plate and causes shattering of the cloves, dry rot due to Botrytis allii, and bacterial rots

Leek

Leek is a non-bulb forming member of onion family. Leek is believed to be originated in the Mediterranean Region.

Growing Leek: A detailed account of various growing practices for leek plants is as given below:

Climatic and Soil Requirements: Climatic and soil requirements for leek growing are same as those of onions.

Propagation: Propagation is via seeds.

Site Preparation: Main site is prepared by ploughing, levelling and by incorporating organic manures and fertilizers into the top soil. Generally, 10 to 20 metric tons of farm yard manure is incorporated into the top soil to enhance soil fertility. One to two month old seedlings are transplanted in trenches prepared in the main field. Each trench is of 30 to 45 centimeters deep.

Nutrient Requirement: Since leek plants have more vigorous growth, nutrient requirement is more than that of onions. Research in this field reveals that 30 tons of leeks per hectare remove 100 kg of nitrogen, 65 kg of phosphate and 130 kg of potash from the soil.

Weeding: Manual weeding is recommended

Irrigation: Irrigation frequency depends on climatic conditions and soil moisture level. It is important that soil moisture needs to be kept at optimum level throughout all growth stages of the onion plant. Irrigation needs to be stopped when plant tops start maturing and showing symptoms of wilting.

Harvesting: Harvesting is done when the plant starts wilting.

Food Uses: Leek is a non-bulb forming member of onion family. The blanched stems and leaves of leek are used as a vegetable. It is used in salads, stews and soups.

Note: More detailed information on onion, garlic, and leek is available in my book titled **"Bulbous Vegetables"**

Celery

Scientific Name of Celery is *Apium graveolens*. Celery is a popular leafy vegetable in many parts of the world and its edible portion is the long, thick, green fleshy petioles and associated leaves. Even though celery is a biennial plant belonging to the Umbelliferae (Apiaceae) family it is commercially grown as an annual crop.

Food Uses: Celery is a popular salad crop. Fleshy petioles and associated leaves of celery are eaten raw as salad. It is also used as a flavouring agent for flavouring vegetable juices and soups. Celery seeds are used as condiment. Celery seeds are also used in many natural medicines.

Production Practices: Various growing practices for celery plants is as given below:

Climatic Requirements: Celery plants thrive well in a cool temperate climate and a humid environment. Celery plants prefer moderate and well-distributed rainfall for healthy vegetative growth. Celery can also be raised as an irrigated crop in dry conditions provided there is a well established irrigational system. However, both extreme and low temperatures are not good for its growth. Extreme temperatures may cause bitterness in celery leaves while the crop bolts at a temperature below 15°C.

Soil Requirements: Celery thrives well on well-drained loamy soils that are rich in organic matter. Celery plants are very sensitive to acid soils; hence pH of 6.6 is considered optimum for their growth.

Celery Varieties: A detailed account of popular celery varieties is given below:

- Florida Golden: Popular variety in the Florida region of USA
- Wright Grove Giant: high-yielding, tall, Green Leaved medium-late variety, producing large, Celery white stalks of fine quality
- Fort Hook Emperor: a late variety having dwarf and stocky, solid, white, thick, broad and tender leaves
- Standard Bearer: An early variety having medium-tall plants, medium pink stems with white longitudinal streaks, sticks solid, good sized

Commercial Production of Celery: Commercial growing practices for celery plants are explained in detail as below:

Land preparation: By ploughing and planking and pulverizing the top soil and leveling the land.

Propagation: By seeds

Seed rate: 100- 200 grams / hectare

Sowing Process: Sowing is done in the nursery beds to raise seedlings. Direct sowing in the main fields with regular irrigations may also be practiced. Sowing time is August–September in tropical regions and March–April in temperate regions. Soil temperature of about 15°–20°C is required for seed germination.

Transplanting age: 4-5 week old seedlings are transplanted in the main field after proper hardening process. Spacing in the main field is 60cm × 20cm.

Fertilizer Schedule: Recommended fertilizer dose is, 200 kg N (nitrogen); 100kg of P_2O_5 (phosphorous) and 150kg K_2O (potash) per hectare for tropics and sub-tropics. At the time of land preparation, a basal dose of 20–25 tons of farmyard manure (FYM) or compost is applied. 1/3 rd N is applied at planting time; another 1/3 rd of N 30–40 days after planting and remaining 1/3 rd of N is applied 15–20 days prior to first harvesting. Full doses of P and K are applied at planting time.

Irrigation Schedule: Frequent irrigations with proper drainage are required for successful celery cultivation. Irrigation should be scheduled in advance and regular irrigations should be done according to the plant requirements. First Irrigation is done immediately after transplanting and subsequent irrigations at 15–20 days interval. Later irrigations are given after every top dressing of nitrogenous fertilizers.

Weed Control: Either manual weed control or chemical weed control may be practiced for celery cultivation. Since it is a long duration crop, weed control is necessary and weeding should be carried out regularly. Biodegradable mulching is also helpful for weed control.

Hoeing: Light hoeing should be practiced regularly; during hoeing process, lateral shoots of the celery plants should be removed.

Blanching: Blanching is a common practice in the commercial production of celery. Blanching is the process by which petioles and associated leaves are protected from direct sunlight in order to prevent the development of chlorophyll in these parts. Major objective of proper blanching is to make the crop crisp; reduce acrid flavours in the leaves and petioles and also to enhance good flavour and tenderness.

Horticultural Maturity Indices: Harvesting stage of celery depends on the market demand that is when the overall field reaches the desired marketable size of 35 to 41 cm (14 to 16 inches) stalk length and before the outer petioles develop "pithiness."Early harvests before the plants reach full size produce high market quality.

Since celery has very uniform crop growth, fields are harvested only once. During harvesting process, tender petioles along with the associated leaves should be harvested. Harvesting time is May in plains and November and April–May in the hills. Expected yield is 350–550 Quintals/Hectare.

Packing: Leaf stalks are packed by size after trimming outer petioles and leaves.

Quality Indices: Major quality indices for celery are the quality of petioles and associated leaves. Petioles and Associated Leaves should be well-formed with thick petioles that are compact, slender and straight. Petioles should be light green in color, and should be fresh. Other Quality Indices that are considered are the crop should be free of any defects i.e. freedom from defects such as bruises, blackheart, pithiness, seed stalks, cracks; and absence of insect damage, yellowing and decay.

Grades: Major U.S. Grades are Extra No. 1 and Extra No. 2. Grade "Unclassified" designates a lot which is not graded.

Packaging: Celery for local markets is normally field packed in 27.2 kg cartons containing 48 stalks. 12.7 kg cartons are used for packing 12 or 18 celery hearts. Celery hearts are prepared from celery petioles that are smaller than regular size. Heart preparation requires trimming of petioles to 20, 25, or 30 cm (8, 10, or 12 in) in length, and then packing in 8 or 13 kg (18 or 28 lb) cartons.

Pre-cooling: Hydro-cooling at 0 °C (32 °F) or Vacuum-cooling at 0 °C (32 °F) is recommended for celery.

Temperature and Relative Humidity: Optimum Temperature is 0°C (32°F) and optimum Relative humidity is 98-100%. At these optimum conditions storage life is up to 5 to 7 weeks.

Optimum Storage Conditions: Trench storage at room temperature or cold storage at 0°C at 95–98% is recommended. Celery should be stored in an isolated place as it easily absorbs other odours. Shelf Life in cold storage is for a period of 2–3 months. Storage-life is reduced to less than 2 weeks at 5 °C (41 °F). At more than 0 °C (32 °F), inner petioles may grow resulting in quality loss. Canning after proper blanching enhances preservation.

Controlled Atmosphere Storage: Reduced O_2 (2% to 4%) and elevated CO_2 (3% to 5%) delay senescence, leaf yellowing, and decay. However, at low O_2 or high CO_2, injuries may occur resulting in off-odours, off-flavours, and internal leaf yellowing.

Retail Outlet Display: For retail outlet display, bundled twist-tied celery stalks are displayed with or without a plastic sleeve. Pre-packaged consumer bags are the best display options for celery hearts. Top icing and misting is used to reduce moisture loss and maintain freshness.

Chilling Sensitivity: Celery is not chilling sensitive. Celery can be stored as cold as possible without freezing. The freezing point for celery is -0.5 °C (31.1 °F).

Ethylene Production and Sensitivity: Ethylene production is low; i.e. < 0.1 μL kg-1 h-1 at 20 °C. The effect of ethylene is temperature and concentration dependent. Celery is not very sensitive to low concentrations of ethylene when exposure occurs at low temperatures. At temperatures above 5 °C, exposure to > 10 μL L-1 ethylene results in loss of green color and development of pithiness in celery stalks.

Physiological Disorders: A detailed account of various physiological disorders of celery plants is given below:

Pithiness: Pithiness is a major source of quality loss and decreased shelf-life in celery. It is characterized by the appearance of whitish regions and air spaces within the tissues and reduced tissue density, and is caused by the breakdown of the internal pith parenchyma tissues of the petiole to produce aerenchyma. Pithiness may be induced by pre-harvest factors, including cold stress, water stress, pre-bolting (seed stalk induction), and root infection. Storage temperature has a major impact on development of pithiness after pre-harvest induction. Progressive development of pithiness is delayed by storage at 0 °C (32 °F).

Blackheart: Blackheart is a physiological disorder caused by cell death resulting from calcium deficiency, and pre-harvest water stress. Internal leaves develop a brown discoloration which eventually becomes deep black. The cause is similar to tip-burn of lettuce or blossom- end rot of tomato. Although many predisposing factors may be involved, water-stress results in a calcium deficiency disorder causing cell death.

Brown check: Brown check is a disorder related to boron deficiency. It appears as cracks on the inner petiole surface and is also referred to as crack stem. The exposed tissues become brown and are susceptible to pathogen infection and decay.

Crushing: Crushing or cracking are signs of mechanical damage, and may lead to cracking rapid browning and decay. Harvesting, packing and handling should be done with great care to prevent damage to the highly sensitive turgid petioles.

Freezing Injury: Freezing injury starts at temperatures below -0.5 °C (31.1 °F). Symptoms of freezing injury include a water soaked appearance on thawing and wilted leaves. Mild freezing causes pitting or short streaks in the petiole which develop a brown discoloration with additional storage.

Diseases: A detailed account of various diseases of celery plants is explained as below:

Bacteria soft rot: Bacteria soft rot is primarily caused by pectobacterium or pseudomonas).

Gray mold: Gray mold is caused by botrytis cinerea.

Watery soft rot: Watery soft rot is caused by Sclerotinia spp. Controlled atmospheres (1.5% o2+ 7.5% co2) have been shown to suppress the growth of Sclerotinia and watery soft rot.

Quarantine issues: There is no quarantine issues associated with the exports of celery leaves. However, export loads of celery may be fumigated at entry ports if common insects (aphids, thrips) are found.

Bibliography

Choudhary, B. (1992). *Vegetables.* New Delhi: National Book Trust of India

Handbook of Agriculture. (2005). New Delhi: ICAR.

UC Davis Post Harvest. (2015, February Monday). Retrieved February Monday, 2015, from UC Davis Post Harvest Technology Center: http://postharvest.ucdavis.edu/pfvegetable/Spinach/

USDA Nutrient Database . (2015, February Monday). Retrieved February Monday, 2015, from USDA Nutrient Database: http://ndb.nal.usda.gov/ndb/search/list

USDA Plant Database . (2015, February Monday). Retrieved February Monday, 2015, from USDA Plant Database: http://plants.usda.gov/core/profile?symbol=spol

ABOUT THE AUTHOR

Roby Jose Ciju is the author of '*The Art of Perfect Living*', an inspirational book based on scriptural wisdom. She is a professional horticulturist and an agribusiness consultant with a Masters Degree in Horticulture and a Post Graduate Diploma in Agri-Supply Chain Management. She has founded www.agrihortico.com, a website dedicated for publishing information on Food & Agriculture Topics. You may follow agrihortico at https://twitter.com/agrihortico1.

Roby has written more than 40 books on various topics till date and her bestselling books are, Mushroom Farming, Moringa, Curryleaf, and Growing Ginger, Turmeric and Arrowroot. Her personal website is available at www.robyjoseciju.com. She may be contacted at roby@agrihortico.com.